PELICAN BOOKS

NUMBERS

Graham Flegg was educated at Wrekin College, St Andrews University, Edinburgh Theological College, and the College of Aeronautics (now Cranfield Institute of Technology). He has been an Education Officer in the RAF, and a lecturer at the University College of North Wales, Bangor, leaving to become a founder-member of the Open University as Reader in Mathematics. There he has developed teaching and research in the History of Mathematics, and he was President of the British Society for the History of Mathematics from December 1979 to December 1982. He has previously written books on Boolean Algebra, Mathematical Structures, and Topology, as well as a number of Open University texts on mathematical subjects and on the History of Mathematics, and has presented Open University broadcasts on radio and television in these areas.

In recent years, his other interests have included theology and music. He is also an international hockey umpire.

NUMBERS
Their History and Meaning

GRAHAM FLEGG

PENGUIN BOOKS

Penguin Books Ltd, Harmondsworth, Middlesex, England
Penguin Books, 40 West 23rd Street, New York, New York 10010, U.S.A.
Penguin Books Australia Ltd, Ringwood, Victoria, Australia
Penguin Books Canada Ltd, 2801 John Street, Markham, Ontario, Canada L3R 1B4
Penguin Books (N.Z.) Ltd, 182–190 Wairau Road, Auckland 10, New Zealand

First published by André Deutsch 1983
Published in Pelican Books 1984

Made and printed in Great Britain by
Richard Clay (The Chaucer Press) Ltd,
Bungay, Suffolk

Contents

Preface

We hear a great deal today about the problem of numeracy. Educationalists vie with each other in their attempts to diagnose the causes of the problem and to suggest ways in which it can be resolved. The blame is laid variously upon parents, primary schools, secondary schools, poor curriculum planning, the examination system, teacher-training, and the universities, each in their turn. Curricula are redesigned, teachers are retrained, innumerable conferences are organized, and yet the problem seems to remain with us. Nevertheless, the majority of people seem to get by in their everyday life; their understanding of numbers and their ability to manipulate them is adequate for their essential needs. On the one hand, therefore, we have a great many educationalists and politicians expressing concern about a serious problem in society, whilst on the other hand most people live their lives as if no problem existed. We may well wonder if the problem is indeed one of substance.

The truth most probably lies somewhere between the two extreme viewpoints. Most people do handle their day-to-day numerical problems adequately. They also handle adequately many problems which they do not themselves recognize as having significant numerical content, yet which in fact involve quite sophisticated numerical ideas. It is, for example, second nature to read the time. People do not need any deep understanding of the mathematics or the technology involved. Equally, people make purchases in the shops and can quite quickly ascertain whether or not any change is correct. We might well wonder, then, just what is the precise nature of the numeracy problem. We can hardly dare to suggest that no such problem exists – there are too many people declaring that it does.

It seems that the numeracy problem has three aspects, which we might somewhat crudely describe as 'high', 'low' and 'in the middle'. The 'high' problem is related to those upon whose basic mathematical ability the commerce and industry of a country depend. It is claimed, and with some justice, that people who are supposed to be mathematicians are no longer able to actually *do* mathematics as well as their predecessors. The blame for this is generally laid at the door of the so-called 'modern' mathematics curricula which are alleged to have taught scholars a great deal *about* mathematics without adequate emphasis upon how to perform calculations. There may well be some truth in this, but the blame should hardly be placed upon the curricula. For the most part these curricula have served their objectives well.

They were designed to redress an imbalance in mathematics teaching toward calculation for its own sake. This was a fair criticism of the traditional approach to mathematics, but balance was not truly restored – the pendulum just swung in the opposite direction. What was needed was the setting of calculation exercises within the context of real-life problems to give them some meaning. This has now been realized and the emphasis has rightly shifted away from the abstract structures of mathematics towards what is termed the 'problem-solving approach'. This is a good thing, provided that the problems are real and not manifestly artificial.

The 'low' problem is one which is rightly the concern of many educationalists. It is realized that pupils in the average-to-bright range will get by even if the teaching which they receive is not particularly inspired. There are adequate pressures from the family and from the school to ensure that these pupils will pass their examinations. There is, however, a serious problem with the apparently less able children, and this cannot just be dismissed by stating that some people are stupid. A great many of these children are not really 'less able' at all – they are certainly not stupid. They are simply children for whom more conventional methods of teaching are unsuitable, and this situation is at its worst when it comes to the question of learning mathematics. It is very important that educationalists continue to give serious consideration to this. They must be prepared to experiment, and adequate funds must be made available to ensure that proper design, conduct and evaluation of such experiments may continue.

The problem 'in the middle' is much more general and hence less easily defined. It can be said to be that, despite all the efforts made by teachers of one generation after another, an organized understanding of the various aspects of numbers is almost universally lacking. It exists only amongst those who have a special interest in mathematical studies. Indeed, there seems to be a positive resistance to almost all the efforts made to persuade people at large that an attempt to learn more about numbers in any organized way is worth the effort. Everyone accepts the necessity for being able to tell the time, read a railway timetable, handle cash efficiently, and the like, but these activities are seen from a purely practical standpoint – they are not thought to be of any conceptual interest, and they are entirely divorced from each other.

Mathematics is unpopular! Its mention is guaranteed to put a stop to social conversation except at a gathering of mathematicians. It is not merely that it is thought largely irrelevant to the necessities of day-to-day life; it is also that there seems to be an unchallenged view that it is both dull and difficult when compared with other areas of study. This seeming irrelevance and the supposed difficulty are not entirely unconnected, and the connecting thread is abstraction. Abstraction has little place in everyday life – pressures to do things are too great and too many to allow time truly to think about things. Thought is largely reserved for the theoretical politicians, economists, and theologians, whose popular image cannot exactly be described as a favourable one.

It is not however universally true that the fascination of numbers is unrecognized except by the professional mathematician. Many laymen have found numerical problems fascinating in their own right. Indeed, a number of significant mathematical discoveries have been made by people who could be labelled as 'amateur' mathematicians. A great deal of fun has been obtained from numbers by a great many people. This fact does not seem, however, to communicate itself to the public at large. Rather than accept that the study of numbers can be a fascinating hobby, the general public tends to put such people into that wide category which has the label 'cranks'. This almost overtly hostile general attitude to mathematics seems to be of comparatively recent origin. In past centuries it was widely accepted that an understanding of, as well as a facility with numbers, is an essential part of general education. This goes back even beyond the days of Ancient Greece, though it can of course be argued that in those days formal education was confined to the privileged few. Today, in advanced societies at least, that privilege is virtually universal, and so we may well seriously enquire as to the source of the widespread seemingly in-built resistance to learning about the ideas underlying numbers when, at the same time, there is an equally widespread acceptance that mathematics plays an essential role in modern technological society. Man now has the knowledge both to travel into space and to destroy the earth as a habitable planet. This knowledge has to a considerable extent depended upon mathematics – in a sense we can say that mathematics and hence numbers hold the key to the future destiny of mankind. This is nothing new. Numbers have always played an essential role in shaping the evolution of society.

This book has been written with an intention of showing that numbers have been at the centre of man's awareness of his surroundings since well before any times of which we have surviving records. It will show that numbers have provided an answer to man's cultural needs at least since any form of organized human society came into being. It will be seen too that, as society has grown and developed, it has never outgrown its dependence upon numbers. Indeed, that dependence is probably never so great as it is today. A second objective has been to collect together many of the ideas and techniques associated with numbers over the centuries so that parents and teachers may be stimulated to experiment with the way in which children are exposed to their mystery and excitement. Uniformity in education is inappropriate in a society made up of individuals with different abilities, different prejudices, different fears, and different interests. Because each schoolchild is an individual, he or she will respond differently to identical educational stimuli. Because each parent and teacher is an individual, he or she will also have different interests and will respond differently to different areas of knowledge. This raises problems in the home and classroom which will always be with us. We can nevertheless learn much from the past. Unfortunately, it is often difficult or at least time-consuming to find out what that past contains.

This book provides information about mankind's past encounters with numbers and how the various problems which they have posed, both practical and theoretical, have been tackled and either solved or evaded. The account has been linked, wherever possible, with aspects of the concept of numbers as they are understood today. This is not, however, a textbook. It is essentially a 'general' and not a 'specialist' work. But, having said this, we would express the hope that, as well as its being of interest to the general reader and the schoolteacher, the serious student will find a few facts and remarks or an organization of the facts which may be stimulating and useful to him also.

It is our firm belief that jargon is one of the curses of the present age. For this reason a great deal of mathematical jargon has been avoided, sometimes perhaps at some expense of clarity and brevity, at least from a purely mathematical standpoint. No apology is being made for this. What the mathematician may at times find a little tedious will, we hope, be helpful for the general reader who could otherwise easily be persuaded to put the book down as not for him – it *is* for him, and for him primarily. Others are asked to be patient with the limitation imposed on mathematical 'terminology' – a polite word for jargon!

Finally, we should express our acknowledgement with much gratitude to all those who have made the publication of this book possible – to Philippa, who typed out the manuscript, to the artists for the excellent sketches which they contributed, to the designers of the cover, to all those who gave permission for the use of photographs and other copyright material, and to the Publishers for organizing and overseeing the whole operation of production and for being so patient when the exigencies of working for the Open University necessitated a delay of ten months in handing over the completed manuscript.

Graham Flegg
The Open University, 1982

Chapter One
Encountering Numbers

This life's first native source.
(Spenser)

O where is the spell that once hung on thy numbers?
Arise in thy beauty . . .
(Irish song)

There is no way in which we can escape from numbers. Numbers are an integral part of everyday life and, as far as we can tell, have always been so. When we wake up in the morning, our first almost instinctive action is to look at our watch or alarm-clock. This immediately confronts us with virtually all the crucial aspects of numbers. We have to locate the numerals to which the hands are pointing or, if we have a digital timepiece, to read a set of numerals. We thereby come face-to-face with written number symbols. If we have an ordinary clock or watch, the numerals may be Roman, I, II, III, . . . or Hindu–Arabic, 1, 2, 3, . . . ; if we have a digital timepiece they will be the latter. When we are asked what the time is, we have to reply in words – 'two o'clock' perhaps. We have now been forced to make use of a number-word, the word 'two' corresponding to the numeral 2. But the word and symbol relate to a specific context, time, and here we face one of the important phenomena of life – the fact that some things are continuous and some things are discrete. This distinction has its counterpart in numbers. We think of time and space as being continuous. Yet, although we may be confident that we know exactly what we mean, the concept of continuity is by no means easy to define. It is almost certain that any precise definition which we attempted to make would not stand up to serious mathematical scrutiny – that is, unless we are specialist mathematicians and have been initiated into the deep mysteries of those numbers which are known technically as 'real'.

When we come to measure time and space, we encounter another problem. In theory at least, if time and space are continuous then whatever we use to measure them must also be continuous if we are to measure them exactly. We can obviously measure them approximately with something that is discrete. Further, we can measure them as accurately as we choose; yet there are several reasons why we cannot measure them exactly. A digital clock, for example, may measure time to a small fraction of a second. Quartz measuring instruments can measure time to an almost unbelievable degree of accuracy. No instrument can, however, measure time exactly. We can understand this in a crude kind of way if we attempt to measure a table with a ruler or tape. We can measure it to the nearest inch – this is obviously only a very crude approximation. We can measure it to the nearest centimetre – this will be a little more accurate. We can measure it to the nearest millimetre, and, if we use a strong magnifying glass, we can perhaps measure it to the nearest tenth of a millimetre. Somewhere we have to stop – we reach the end of the capability of the measuring system adopted. In any case, there will have been little point in attempting to measure to fractions of a millimetre because our reading also depends on how accurately we have aligned the zero of the measure to the other side of the table. This is a problem which affects all measurement. Although measurement seems as if it ought to be a continuous phenomenon, in practice it is discrete. Again, we find this is reflected in numbers. Just as there are numbers which correspond to the continuous, there are also numbers corresponding to the discrete, the most obvious of these being the whole numbers, though the numbers which correspond most closely to the measuring situation are known as the 'rational' numbers – the ratios of whole numbers.

The problems associated with measurement do not end here, however. If we construct a square whose side measures one unit, there is no way in which we can measure its diagonal exactly using just the rational numbers. No matter how we argue that by dividing up that one unit indefinitely into ever smaller and smaller fractions we ought in the end to be able to measure the diagonal exactly, any reasonably competent mathematician can prove that our argument is false. This does not affect us in practice, however, because we can still measure the diagonal as accurately as is necessary. Mathematicians respond to this situation, known since the times of the Pythagoreans, by classifying the square root of two as an irrational number. Yet, even if we add to the rational numbers all those of the same kind as the square root of two, we still do not have enough numbers to measure continuous quantities exactly, even in theory.

The most fundamental way in which numbers arise is through counting. This is the way in which numbers were first understood by man. When we count, we assign objects to as many of the positive whole numbers as are necessary in turn until we have exhausted whatever it is we are counting. We do this automatically. But this seemingly simple process conceals another important aspect of numbers – the fact that they are ordered. In order to count up to five (say), we not only need to be familiar with the first five number-words, we need to be familiar with them in their order. It is probable that it was this 'ordinal' aspect of counting, as it is called, that was the first aspect of numbers of which man became aware.

Many of the apparent problems which are associated with counting, measuring, and continuous aspects of number arise because of the difference between what we call the 'real world' and the conceptual world which we create in our own minds. This is highly relevant to the question 'what are numbers?' Numbers are the basis on which the whole structure of mathematics has been built, and the positive whole numbers are the basis of all other numbers – for this reason they are called 'natural' numbers. Provided that we accept these numbers as being 'naturally' given, all others can be constructed from them, though it needs some fairly sophisticated concepts to make the jump from the discrete to the continuous and thus construct the real numbers from the rational numbers. This does not however tell us what kind of things numbers are. We get a possible clue if we think about the disparity between measuring in theory and measuring in practice. When we say 'in practice', we mean that we are speaking of something which we can actually do physically. When we say 'in theory', we are speaking of something which we can do in our minds. The distinction is that between physical 'reality' and conceptual 'reality'.

If we take the number two, we can be aware of this number in at least four different ways – as a numeral, as a number-word, as a concept in our minds, and as a property possessed by every collection of two objects. Although for many practical purposes we do not need to worry about these different aspects of numbers, it is very important that we are aware of them in any study of the history of numbers. There is nothing in the physical world which *is* two. There are, however, a great many things in the physical world to which 'two' may be usefully applied. Numbers are thus essentially concepts, and mathematics is the study of these concepts and of the structures which can be built from them. The concept of numbers arises directly out of our experience of the physical world in the same sort of way as our concept of colours. Numbers are idealizations in the mind of particular experiences encountered in the world. The number two does not have an independent existence of its own except as a concept, neither does redness have an existence except as a concept. Perhaps we should have coined the word 'twoness' rather than the word 'two' – the analogy would then have been a little more obvious.

To some extent we are led away from the appreciation of the conceptual status of numbers by the symbols which represent them. We are so used to manipulating these symbols that they come to take on, as it were, a life of their own. Since the symbols clearly exist in the physical world, we tend to grant the same status of existence to the concepts which they represent. The fact that we encounter and use both numerals and number-words every day of our lives gives them a deceptive familiarity. They exert their own particular spell upon us, and the concepts which they represent are thereby translated from the realm of the mind into the physical world which surrounds us.

Numbers are, nevertheless, endemic to the natural world in some remarkable ways. If numbers are conceptualizations of man, we may well ask how it is that they appear to govern so many of the phenomena in nature. Some of the most beautiful shapes to be seen in nature are found on close examination to be governed by series of numbers. This indicates that numbers are in some way connected with aesthetics, a connection often exploited by artists and architects in ways which very much correspond with the harmonies of music. The centre of a daisy is composed of scores of tiny florets arranged in two opposite sets of spirals, 21 spirals in a clockwise direction and 34 in a counter-clockwise direction. A pineapple has eight spirals of bumps going in the clockwise direction and thirteen in the counter-clockwise direction. Pine-cones are built up in a similar pattern; this time five spirals go clockwise and eight counter-clockwise. There is nothing very spectacular in all this – not until we begin to look for other connections between these numbers.

If we start with the number one, and create a sequence of numbers built up in such a way that each number is the sum of the previous two numbers, we obtain:

1, 1, 2, 3, 5, 8, 13, 21, 34 . . . .

This sequence includes all the pairs of numbers which we have just noted to exist in various ways in nature. The special form of this sequence, known as the 'Fibonacci sequence' after its discoverer (Leonardo of Pisa, son of Bonaccio) has been known since the twelfth century. It arises directly from the following problem:

> How many pairs of rabbits can be produced from a single pair in a year if every month each pair begets a new pair which, in turn, becomes productive from the second month onwards?

Again, this may not seem particularly interesting or extraordinary. This sequence is, however, by no means confined to daisies, fruit and a problem about rabbits. It occurs again and again in nature. It occurs, for example, in the way in which successive leaves grow around the stems of many plants and trees, and it is only one of many such instances in which sequences of numbers, apparently invented by man, are found to have existed in nature since the times before man appeared on the earth. This raises some interesting problems for debate. For example, do we invent mathematics or discover mathematics? Are numbers in some sense at least 'given' so that there comes a point at which any attempt to find still more basic definitions is useless? To what extent are the numbers which we find so extensively in nature, and the relations between these numbers, directing the way in which nature evolves? These and other similar questions have been debated by mathematicians, philosophers, and even theologians over the centuries. The question has been asked: 'is God a number?' Certainly, there has often been a close relationship between numbers and religious beliefs. Some theologians have denied that numbers are creations at all, and have suggested that numbers control both the Deity and His created universe. We shall not debate these questions, interesting though they certainly are. The inclusion of the word 'meaning' in the sub-title of the book is not meant to refer to abstract philosophical questions about numbers but rather to their practical significance in mathematics and in life generally. In the following chapters we are concerned to present as many of the basic facts about numbers as are readily accessible to the layman, and to do so in a historical context.

Most people probably agree that numbers play an important role in everyday life and are crucial to man's economic, scientific and technological development. It is not as fully appreciated that they play just as crucial a role in the 'humanities', even though this was well-known in Ancient Greece. It comes, of course, as no surprise to the mathematicians. No one who is truly versed in the art of numbers and the structures which can be built from them can fail to be aware that they have a particular kind of beauty which is all their own. The popular image of the pure mathematician as one who is divorced from both the harsh realities and the aesthetic beauties of the world lies far from the truth. In fact, the great majority of mathematicians have a serious interest in aesthetic studies of one kind or another. This is not because such studies are a relaxing contrast to mathematics, but because beauty of shape and sound is felt to be a reflection of the beauty and order to be found in numbers.

It has been claimed that mathematics is the 'queen of the sciences'. This carries with it the suggestion that mathematics has feminine characteristics, yet it was first made at a time when practical activities were thought of as being essentially masculine and the arts feminine. We might also note that justice and mercy were thought to have respectively masculine and feminine characteristics. In a sense, mathematics – and numbers in particular – effect a marriage between the complementary masculine and feminine aspects of life. Numbers reveal the unity which underlies all of life as we experience it. There is an increasing awareness of this today, despite the widespread popular prejudice against mathematics. We are perhaps gradually returning to the viewpoint of the Greek philosopher mathematicians whose belief was that 'all things are number', re-echoed in the claim of the nineteenth-century French philosopher Auguste Comte that 'there is no enquiry which cannot finally be reduced to a question of numbers.'

Chapter Two
Counting with Numbers

The King was in his counting house
Counting out his money.

(Nursery rhyme)

He counted them at break of day –
And when the sun set where were they?

(Lord Byron)

Counting is an everyday activity of man, and has been since before the dawn of history. It is an activity which is inseparable from speech. To be able to count, we must know a sequence of number-words and be able to relate these in their proper order to whatever is being counted. This does not mean, however, that an abstract understanding of numbers is needed. The intellectual step taking us from counting to numbers in the abstract is a comparatively sophisticated one which came late in man's history. Counting was the first of a long succession of practical and intellectual steps which has led to the mathematics of today. It lies at the root of all that we have learned about numbers from the simplest arithmetic to the complex calculations which have enabled man to set foot on the Moon and to devise the means of his own total annihilation.

Rudimentary Beginnings

There are certain rudimentary senses which man shares with many other creatures. These include an awareness of size and shape and some sort of an appreciation of quantity. Without them, counting would be impossible.

Man's sense of size enables him to distinguish between one object and another. It enables him to appreciate, for example, the special kind of difference between a pebble and a boulder or between a mountain and a hillock. This sense of size has always been a necessary part of his reaction to the world about him. A basic sense of size precedes any development of the concept of numbers. Numbers do not become involved even implicitly until the need arises to consider the result of putting, actually or mentally, several objects in order according to their sizes. Numbers also become unavoidable as soon as man needs to measure, however crudely, and to express differences in the sizes of various objects in quantitative terms.

Man's sense of quantity gives him the ability to see that there is a particular kind of difference between one collection of objects and another. It enables him to see that, for example, his neighbour has more cattle than he, and to do this long before he is able to associate it with numbers.

In the first instance these two senses will have been applied only crudely. This means that there will have been little difficulty in appreciating size differences between objects of very different kinds. There is, for example, little distinction between the awareness that a mountain is larger than a hillock and the awareness that it is larger than a stone. But when the sense of size becomes more refined and objects more nearly equal in size have to be compared, differences in shape become crucial. Eventually, with objects of the same kind and shape, it becomes ever more difficult to be sure about relative size, and we arrive at the need to carry out some kind of comparative measurement.

In a similar way, with the sense of quantity, crude comparisons between collections of large numbers of objects and those of much smaller numbers of objects present no difficulty. This is true even when the objects themselves are of different kinds. As the sense of quantity is applied in a more refined way, comparison becomes increasingly difficult. Eventually the need for more advanced abilities cannot be avoided.

We know from observation and experiment that man shares the basic senses of size, shape and quantity with many other creatures. We must beware, however, of reading too much into what we observe. Schoolboys know that one egg can be taken from a nest apparently without the mother bird noticing anything on her return. Removal of two eggs is almost invariably noticed and leads to desertion of the nest. There is the well-known story of the attempt to shoot a crow which had made her nest in the watch-tower of an estate. When the owner entered the tower she would fly away, returning as soon as he reappeared outside. To try to deceive the bird, the owner entered the tower with one of his men who later reappeared leaving him waiting inside. She was not, however, to be fooled by this simple trick. The next day, the owner took two men inside the tower and remained behind after the two had returned outside. Again, the bird was not caught out. The trick was repeated with three men accompanying the owner, but equally without success. It was only when it was repeated with four men accompanying him into the tower that the bird was deceived. When the four reappeared she returned to her nest and met her fate. Apparently, she could distinguish up to but not beyond a count of four.

These and other similar stories have led some people to infer wrongly that creatures other than man can actually count. In fact, the stories provide evidence only of the rudimentary sense of quantity which in man alone is a prelude to counting. Even the dogs who have appeared on television barking prescribed numbers of times provide no evidence of an ability to count. There are many much more remarkable phenomena in nature, especially amongst the insects, but none of these amount to evidence of counting. They reveal only inborn instincts, carried and refined by the genes of successive generations, or, in the case of the barking dogs, the result of careful conditioning. To the question 'can animals count?', we are bound to reply a firm negative.

These basic senses of size and quantity can also be observed in young children along with the sense of shape. Here, there is clear evidence of appreciation of what they will eventually call 'larger', 'smaller', 'more', or 'less'. The case of the child who takes the larger of two coins irrespective of denomination is a case in point, as also is the choice of a box containing three coins in preference to one containing two. Possession of these inherited senses is essential before there can be any appreciation of numbers or even the most limited ability to count. There are two further abilities required before counting can begin. One of these is the ability to compare two distinct collections of objects by the method of one-to-one correspondence.

One-to-one Correspondence and Order

The ability to compare collections of objects is one of the things which make man unique amongst the creatures. It is the first step toward counting and numbers after this basic sense of quantity. It enables man to distinguish between different collections without giving a specific answer to questions about 'how many'. Such questions, however, became more and more important to man as he settled down in communities and become a farmer instead of being mainly nomadic. He could by now compare, for example, his sheep or cattle with his fingers and toes, one by one, and so find out if all his flock were remaining safe in his keeping. So, in the morning a man who had fifteen sheep (say) would see that his flock corresponded to two hands and one foot. If he made a similar comparison later in the day and found that he had more fingers and toes on his two hands and one foot than he had sheep, he would know that his flock was no longer complete even when he was yet unable actually to count the number missing. Of course, in practice many shepherds were able to identify the members of their flocks individually. This is true even today. Just as we know if someone is absent from a family gathering because we miss them as individuals, a herdsman can identify one or more missing animals because they too are missed as individuals. St. John's Gospel reminds us that the Good Shepherd 'calleth his own sheep by name'. No process of counting need be involved.

It was natural that man should establish one-to-one correspondences with parts of his own body, though it is possible that he also matched things with the members of his family. His body and his family provided ready made collections for comparison. It is, however, only a short step from one-to-one correspondences with collections already there to correspondences with collections specially put together. This becomes necessary when the numbers of things involved are relatively large or when any kind of record is needed. So man began to make marks in the earth or sand, and to make heaps of twigs, shells, or pebbles which corresponded one-to-one with his sheep, his bags of grain, and his other possessions. All this took him one further stage towards an understanding of numbers, but it still did not amount to counting. Nevertheless the use of model as opposed to natural collections for establishing one-to-one correspondences represents a further step towards the abstract concept of quantity.

The second major ability required to complete the prelude to counting is the awareness of order. Fortunately, there is a natural order in the parts of the body and especially in the fingers and toes which not only suggests the idea of 'next in sequence' but also eventually provides the basis for spoken words for numbers. It is the use of parts of the body for matching by one-to-one correspondence that is the key which opens the door to counting. There is no known example of a people able to count, who have no history of an earlier stage of body or finger matching.

Order arises naturally out of the process of addition by ones. We must start with one finger, one pebble or shell, one mark scratched in the sand, or one notch cut in a stick, and then add another to what we already have. We continue the process by adding one more, then another, and so on. Given, for example, a heap consisting of three pebbles, what must inevitably follow in order of succession is a heap consisting of four pebbles. What comes next is obtained by adding just one pebble to those that are already there. The same applies when fingers are used, whether it is the thumb or the little finger which begins the sequence. A specific order is established without which counting is impossible. There are, however, two essential differences between using pebbles, shells, scratch-marks or notches and using parts of the body. If we are presented with a group of (say) four pebbles, what is suggested to us numerically is just the total four. The idea of a sequence is not visually apparent. If we hold up four fingers on one hand, however, we have a visual representation of both the total four and the sequence by which it was reached. This is so because the fingers are clearly distinguishable from each other and are in a fixed order. It is this fact of their being clearly distinguishable and hence having special names, 'little finger', 'ring finger', and so on in a natural sequence, which finally makes counting possible.

No doubt, man had some kind of awareness of order before he invented or adapted the words with which he could describe the number of objects in a collection or express the position of any particular object in a sequence. Some glimmer of appreciation of numbers is thus possible even when no number-words or specific number-symbols have been invented. At this stage, however, everything has to be conceived pictorially. There is no understanding at all of numbers in the abstract, and counting as such has not quite been achieved. Nevertheless, there has been considerable progress beyond the capabilities of other creatures, and much that is useful for everyday life.

We do not know just how or where man first came to appreciate the concepts of one-to-one correspondence and order. There is no direct evidence remaining from this early period of man's intellectual development. Even the order in which the various senses and abilities came to fruition is a matter for speculation. The best that we can do is to suggest that it seems most likely that at certain times, lost to us in the mists of pre-history, there were various stages of intellectual and practical development which eventually led man to an appreciation of number and the ability to count. This development led him from the awareness of things being equal or unequal in number, applied first to objects of the same kind and later to objects of different kinds, through the idea of one-to-one correspondence and an appreciation of order to counting. We can assert that matching objects with various parts of the body, and especially with the fingers, played a crucial role. Counting itself is thus an advanced process; it is by no means as instinctive and innate a process as it might appear to be.

Some writers claim that we can find out how man's understanding of number came about by watching the behaviour of young children. This kind of evidence is valuable for learning how number comes to be understood by today's children, but the historian has to treat it with great caution because the environment in which children grow and learn today is very different from that in which man first learned to count. Perhaps of more value for the historian is the evidence gained by studying undeveloped tribes still living in various continents. But even this can mislead. It is valuable for learning about different kinds of counting systems, but it is doubtful if it throws much light on the earliest stages of man's intellectual development.

The Beginning of Numbers

At some time or other, along with the development of speech, individual words or phrases came to be used to describe different numbers of objects. These were first directly related to parts of the body used for one-to-one correspondences or to gestures associated with specific quantities. We infer this from investigations of the number-words of many undeveloped tribes. Although it is comparatively rare to find the actual names of fingers used directly as number-words, phrases describing raising or, especially, bending fingers abound. A typical example would give a sequence of number-phrases like:

the end one is bent
another is bent
the middle one is bent
one is still left
the hand has died.

Here, the process of counting involves direct correspondence between objects and fingers; there is no requirement for any awareness of numbers in the abstract.

Even after number-words had become detached from the parts of the body or gestures associated with counting, they did not immediately become fully independent. There is much evidence from linguistics to suggest that for a long time they were directly associated with the particular objects being counted. At this stage, three sheep could be distinguished from two sheep and three fingers from two fingers, but the abstract appreciation of three and two was yet to be understood. Even after he had devised effective methods of counting, it was a long time before awareness of number in the abstract was achieved. Going from the association of three fingers with three sacks of corn and with three loaves of bread to an appreciation of threeness and hence three in the abstract involved an intellectual leap of considerable proportions. Adding two fingers and three fingers to make five fingers corresponds to adding three sheep and two sheep to get five sheep; this was appreciated, but only because in both cases there was a corresponding physical action or event in everyday life which gave the addition meaning.

It is possible that there was a kind of half-way-house on the road to abstraction when counting and addition of objects were performed solely in the mind. Mountains, for example, could not be moved about so that they were in a convenient group together. Counting of mountains would thus have to involve an act of the imagination, as would numbering objects in relation to future hopes and plans. Three wives could be conceived of and counted mentally by an aspiring bachelor and then discussed with his friends. Such mental counting may have been the door to the abstraction of number itself where three and two always give five, whether it be sheep, mountains, future wives, or fingers on the hand. This stage, however, came very late.

The linguistic evidence is provided by present-day languages, including those spoken by undeveloped tribes, and written documents surviving from the past. This means that it is evidence of only recent origin when compared with the overall development of man's understanding of number. Even when we use linguistic methods in an attempt to recreate languages, the details of which are now lost to us, we can go back only a few millennia. Such evidence does, however, suggest that the abstract concept of number presented a real intellectual hurdle to man.

Examination of many number-words suggests just how closely numbers were tied to objects being counted even after quite efficient counting systems had been invented and the ability to carry out simple calculations had been developed. Many special words exist in English for collections of particular kinds. We think of words such as *flock*, *herd*, *bunch*, *shoal*, *litter*, and so on, which we apply to specific things. It would not be correct to speak of a herd of grapes, or a litter of fish! This applies also to words in many languages representing collections of just two things, such as *pair*, *duet*, *yoke*, *brace*, in English. We would hardly speak of a brace of shoes! An extreme case of this is an Indian tribe in British Columbia which at one time had as many as seven different sequences of number-words dependent on what was being counted.

This kind of evidence is re-inforced by the survival in a number of languages of number-words which change in some way according to the objects to which they are being related. Thus we find *un* and *une* in French both corresponding to English *one*; they are directly related by gender to what is being counted. A similar situation is to be found with Italian and Spanish. Sometimes there are three such forms of number-words, masculine, feminine and neuter, as with Gothic *twai*, *twos*, *twa*, all equivalent to English *two*. This can be found also in some Slav languages. Some number-words are also affected by case. Many of us will have had to learn to decline the Latin words *unus*, *duo* and *tres* for English *one*, *two* and *three*. In all these examples, the number-words are treated as adjectives, so it would appear that numbers were originally thought of as attributes of what was being counted.

Numbers are, however, also linked with objects by changes in the nouns by which the objects are represented. In English, we speak of one *man* but of two or more *men*, of one *book*, but of many *books*, and so on. In certain languages in addition to distinct singular and plural forms of nouns, we find a third form used in the case when exactly two things are being spoken of. This third form, known as the dual, is found in classical Greek, Biblical Hebrew, Arabic, and elsewhere. We have Arabic *radjulun*, *radjulani*, *ridjalun*, used when speaking respectively of one man, two men, and more than two men. Examples also exist, though they are very rare, of distinct forms of nouns when three and even four objects are being referred to.

Both the adjectival forms of number-words and the changes in noun forms dependent on number indicate, as we have already said, that counting was possible without there being any abstract concept of numbers. This must represent, however, an early stage in the development of counting. The surviving examples of special associations between number-words and things counted do not extend to numbers greater than four. This suggests that at one time four was a kind of universal limit to counting and that the intellectual leap to numbers in the abstract did not take place until that limit had been overtaken. Number-words beyond four may have been invented well before the number sequence became detached from the objects counted. We do not, however, have the concrete evidence needed to establish exactly when this intellectual leap took place.

Patterns and Groups

Counting by one-to-one correspondence with parts of the body or with pebbles or shells rapidly gives rise to difficulties when large numbers are involved. Large herds of animals, plentiful sacks of provisions, boxes full of many pebbles, and so on, all point to the necessity for some sort of grouping of numbers if the counting process is to remain manageable. So man's sense of shape was brought into use to assist him in developing the counting process to meet the ever increasing need to take account of large numbers of objects. We may well speculate just how far counting can be continued without any assistance being obtained from repeating patterns or from grouping objects in a number of equal collections. We are not very good at estimating the number of objects present in a large randomly arranged collection. For this reason, guessing the number of sweets in a glass jar is a common money-raising feature at bazaars and garden parties. Some less developed peoples do seem, however, to have a highly proficient skill in estimating large numbers. The Kpelles of Liberia, for example, are reputed to be experts at visual estimation of the number of stones in a pile as well as of the amount of cereal in a container. Yet, curiously enough, they are also said not to be skilled in counting larger objects such as the houses in a village.

The use of patterns and groups to assist with the counting process is now a universal practice. We find it in our own common habit of totting up in fives – after drawing four strokes IIII, we form a distinct group to represent five by drawing the fifth stroke through the previous four to give 卌. In a similar manner, the bank cashier will count pound notes by bundling them together in tens or twenties, or even larger bundles. After each group or bundle is completed, counting can start again from the beginning of the number sequence. Some groups suggest themselves automatically. In particular, the use of fingers and toes for counting immediately suggests a natural grouping of five.

Sometimes, counting is assisted by placing objects in patterns which have a particular cult significance. The Pueblo Indians of South America, for example, arrange pebbles, sticks and arrows in a variety of ways, different patterns representing various different numbers.

In a modern army, a count of the number of men drawn up in a platoon is obtained by numbering the front rank only; it is assumed that the ranks behind are correctly covered off. The idea of determining the number in a group by matching with a similar group of known count is clearly of ancient origin. We can think of this, if we like, as group-by-group correspondence. Herodotus, for example, recounts that Xerxes counted the number of men in his army by first packing a known number of men as closely together as possible and then building a low wall around them. The rest of the army was counted by repeatedly packing the enclosed space with new men until all were accounted for.

Even when there is a suitable sequence of independent number-words available, it has to be limited by the demand which it makes on the memory. This limitation is achieved by the use of group words. This in turn leads to the invention of systems for combining number-words in various ways so that large numbers are represented by a combination of words for smaller numbers. Our English number-word *thirty-five* simply derives from the expression 'three tens and five'.

Once this stage of forming combinations of number-words has been reached, we have arrived at a counting system – the culmination of a long process of intellectual development. Many different counting systems have been invented in different places and at different times. Today, apart from the specialized needs of the electronic computer, there is virtually universal acceptance of decimal counting systems, such as the system of English decimal number-words. Remnants of non-decimal systems do still exist, however. These are to be found only amongst the indigenous tribes largely untouched by modern civilization, in gradually disappearing non-metric units of mensuration, and in the forms of certain special number-words actually used in decimal counting systems.

The development of counting systems thus involved both patterns and groups, as the study of the systematic build-up of the various number-word sequences shows. Visual groups were represented by special group words and the sequences were advanced according to clearly recognizable word patterns. Thus, the sequence of number-words *thirty*, *thirty-one*, *thirty-two*, . . . , has exactly the same patterns as the sequence *twenty*, *twenty-one*, *twenty-two*, . . . , which precedes it.

Finger-counting

The origins of finger-counting lie deep in pre-history, and we cannot be certain how or where some of the particular forms of it known to us actually originated. It is perhaps surprising that some of these survive to the present day despite the advent of efficient systems of number-words and written number-symbols. They are still to be seen as a common feature of the market places and ports of Africa and the Indian sub-continent. They have survived in Europe and America in travelling fairgrounds, and have been developed with some sophistication for use at the racecourse and in the stock exchanges. Our use of the word *digits* to mean 'written numerals' derives from the Latin word *digiti*, almost universally used in Europe in the Middle Ages, and originally having the meaning 'fingers'. The evidence of finger-counting is so widespread that we are forced to conclude that it has been a universal practice.

Various different conventions in finger-counting have been and still are used in different parts of the world. Sometimes, as for example in Africa, differences in convention are so localized that it is possible to distinguish one tribe from another simply by the way in which the fingers are used to denote the number sequence.

Often, finger gestures are used in conjunction with entirely unrelated number-words, with which they may have an equal status. Their great advantage lies in the ability to cross language barriers, though of course the particular convention being used must be understood by both parties to a dialogue. The use of finger-counting, as we encounter it today, does not therefore necessarily imply that an abstract understanding of number is wanting. Some anthropologists have quite unjustifiably jumped to this conclusion and, as a result, have seriously underestimated the numerical ability of the peoples whom they have been studying.

Variations in the conventions for finger-counting include beginning with the little finger, beginning with the left hand or with the right, starting with the hand open and closing fingers one by one, starting with the hand closed and raising the fingers, and so on. For example, there is an interesting account of how, during the 1939–45 war, a Japanese girl passing herself off as Chinese was identified by the fact that she counted on her fingers the Japanese way. The Japanese begin with an open hand and close the fingers; the Chinese begin with a closed hand and open the fingers.

The form of finger-counting common in Europe until its replacement in the sixteenth century by written Hindu–Arabic numerals was probably of Roman origin. It had the advantage of being fairly easily learned, even by the uneducated, and it was particularly useful in affairs of commerce because it trans-

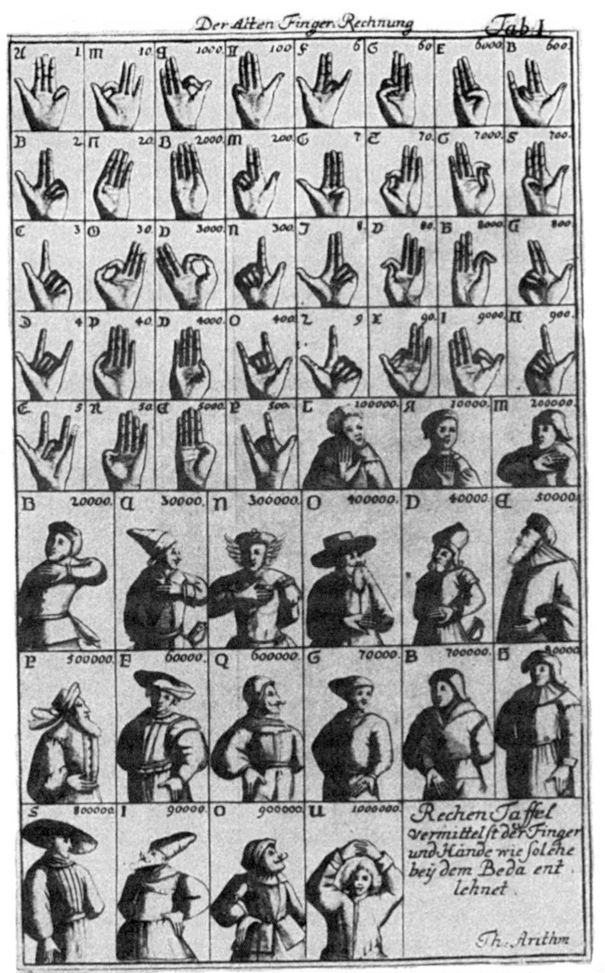

Bede's finger counting (Jacob Leupold, 1727)

cended language barriers. Finger gestures were commonly used in association with calculation on the abacus, usually for keeping a temporary record of the results obtained at different stages of a calculation. But, even after the advent of modern numerals, they continued to be used.

There are many surviving references to this finger-counting in Europe, one of the most famous accounts being that of the Venerable Bede. Bede, who died in 735, left us full details in an introduction to a work dealing with the computing of the date of Easter. This was clearly a most important treatise and we find reference to it and descriptions of its system of finger gestures in many works, including the famous *Summa* of Luca Pacioli published in 1494, and in German works dating from as late as the eighteenth century. The accounts are not always completely consistent: Pacioli's, for example, reverses the gestures used for hundreds and thousands, as does Leupold's.

In addition to the various detailed descriptions of finger-counting gestures, there is much scattered evidence of various kinds. Finger-numbers are depicted on Roman counters dating from the first century A.D. Some early commentators on Biblical texts, St. Jerome and St. Augustine of Hippo for example, both of whom died in the third decade of the fifth century, clearly assume a knowledge of finger-counting on the part of their readers. They both attempt to see mystical significance in numbers and derive this from their association with the fingers. As an example, we find that St. Jerome, commenting on the Parable of the Sower in St. Matthew's Gospel, states that the bringing forth of the fruit 'thirty-fold' symbolizes marriage because the way in which the index finger and thumb of the left hand are placed to denote thirty represents the joining together of man and wife. 'Sixty-fold' is said to symbolize widowhood because the gesture in which the thumb presses down on the index finger to denote sixty represents the weight of the tribulation which widowhood bears. 'Hundred-fold' is said to be symbolic of virginity because hundred is represented in a similar way to thirty but on the other hand.

Symbolic interpretation of finger-numbers was by no means confined to commentaries on Christian writings. There are references in writers such as Pliny to the finger gestures to be seen on statues of pagan gods. A statue of Janus, for example, is said to identify him as the god of the year and hence of the sun because it shows him forming the number three-hundred-and-sixty-five with his fingers. We can find references also in Byzantine and in Arab works. In the case of Byzantine icons, however, we need to be careful. Some writers have mistakenly interpreted finger gestures forming Greek letters IC and XC (the first and last letters of 'Jesus' and 'Christ' in Greek) as representing numbers.

The system described by Bede begins with the little finger of the left hand. This is bent at the middle knuckle with the tip touching the palm of the hand to represent one. The sequence continues:

two:	bend down the ring finger
three:	bend down the middle finger also
four:	raise the little finger
five:	raise the ring finger
six:	raise the middle finger and bend down the ring finger
seven:	raise the ring finger and bend down the little finger at the bottom joint
eight:	bend down the ring finger at the bottom joint also
nine:	bend down the middle finger at the bottom joint also

The numbers one to nine are therefore represented using only three fingers of the left hand. The tens are also represented on the left hand, the sequence being:

ten:	put the nail of the index finger against the middle of the thumb
twenty:	put the tip of the thumb between the index and middle fingers
thirty:	make a circle with the thumb and index finger
forty:	put the thumb on top of the index finger with both extended

fifty: bend the thumb across the palm
sixty: put the tip of the index finger over the thumb (still bent across the palm)
seventy: bend the index finger and put the thumb nail inside the bend
eighty: straighten the thumb inside the bent index finger
ninety: put the tip of the index finger on the inside of the base joint of the thumb

Tens up to ninety are thus represented using only the thumb and index finger of the left hand.

Hundreds are now represented on the right hand in the same way as the tens on the left, and the thousands on the right hand in the same way as the units on the left. It is therefore possible to represent any number up to nine-hundred-and-ninety-nine by combinations of the various finger gestures described. Eleven, for example, is indicated by putting the nail of the index finger of the left hand against the middle of the thumb (ten) and at the same time bending down the little finger at the knuckle. If we hold our hands up with the palms facing a mirror, remembering that our hands are reversed in the mirror image, we can see that we have in effect a kind of place-value system in which numbers are read in ascending order from right to left.

Bede goes on to describe finger gestures for numbers up to one-million, but it is doubtful if these were ever in general use. Certainly, these additional gestures were of comparatively late origin; the Romans used finger-counting only up to ten-thousand.

Numbers were divided into *digiti* (one to nine), *articuli* (numbers divisible by ten), and *numeri compositi* (numbers above nine not divisible by ten). We find this division in the sixth-century writings of Boëthius. It is directly related to finger-counting since *digiti* is Latin for fingers and *articuli* is Latin for joints, that is, knuckles.

It is unfortunate that Pacioli chooses to reverse the representation of hundreds and thousands because this destroys the place order. Changes made in descriptions by a number of other writers in the sixteenth century and later are clearly just incorrect; they destroy the logical order of successive gestures. A knowledge of the correct sequence enables us to interpret a number of otherwise seemingly obscure phrases in the works of many authors. For example, there was a fairly well-known riddle, quoted by Alcuin but dating from Roman times, which went as follows: 'a man holds eight; from this he takes away seven and six remain'. To understand this, we must first bend down the ring and little fingers of the left hand at the bottom joint, thus forming eight. Then, we must raise the little finger, which by itself represents seven, and we find ourselves with just the ring finger bent down. This denotes six. Again, the second-century Roman writer Juvenal refers to a very old man as 'one who has cheated death over the years and now counts his age on his right hand'. This means that the man has reached the ripe old age of one-hundred years. Even the writers of training manuals on hunting and war assumed familiarity with finger-counting. In a thirteenth-century book on falconry, written by the Emperor Frederick II, the way in which the fingers should be arranged when holding a falcon is minutely described in terms of representing the number seventy-three. In another work, instructions on handling a bow prescribe that the string should be held with the fingers forming the number thirty.

The finger-counting conventions to be found today in markets, racecourses and exchanges differ substantially from those described by the Venerable Bede. In the United States, elaborate gestures have evolved which enable buyers and sellers to communicate bids and acceptances with almost incredible speed and accuracy. Accuracy is possible to fractions of a cent. In ports and markets in the Near and Far East, a remarkable system of finger gestures, performed in secret under a cloth, has evolved. Here, numbers are read off by touch.

Many different systems of finger-counting are to be found in the continent of Africa. Some must be of great antiquity. Specific references to counting on the fingers date back to at least the sixteenth century B.C. In *The Book of the Dead*, for example, a work devoted to the passage to the other world which appears increasingly in the more notable Egyptian tombs from the time of the New Kingdom, it is related how, in order to cross the river of death, a deceased king is required to recite a rhyme which numbers each of his ten fingers. We do not know, however, precisely how fingers were related to numbers in this case.

In twentieth-century Africa there are two main types of convention to be found. One is the obvious one of raising fingers successively on one hand and then on the other, and the other is the more unusual convention of representing numbers greater than five by raising an equal or as nearly an equal numbers of fingers on each hand as is possible. Zulu finger-counting is an example of the former. The gestures proceed:

one: raise left little finger
two: raise left little and ring fingers
... ...
five: raise all left fingers and thumb
six: raise right thumb
seven: raise right thumb and index finger
... ...
ten: raise right thumb and all four fingers

As a contrast we have the finger-count to be found in North-East Tanzania which proceeds:

one: raise the right index finger
two: raise and spread apart the right index and middle fingers
three: raise and spread out the right middle, ring and little fingers
four: raise the four right fingers with a gap between the middle and ring fingers
five: raise a clenched right fist
six: raise and spread out the middle, ring and little fingers of each hand
seven: indicate four on the right hand and three on the left

eight: indicate four on each hand
nine: indicate five on the right hand and four on the left
ten: raise right and left clenched fists

The interesting thing about this and the few similar systems is that the obvious gesture of the whole fist for five is not retained for representing six to eight. Just why this variation exists in certain limited areas only is not properly understood. There are also a few very rare examples where both forms of convention are in use in the same area, the one chosen at any particular time depending on what is being counted.

We come across the occasional example of finger-counting on one hand only, usually the right hand. This choice of hand is somewhat surprising as it is the right hand that is most frequently used to carry a weapon and there is no suggestion that right-hand counters are otherwise left-handed.

These various contemporary finger-counting conventions, and their many variants have parallels in the different systems of number-words which we are about to discuss. The widespread evidence which they provide, along with documents, pictures and statues from the past, and the evidence from the number-words in many languages, combine to confirm how universal has been the practice of using the fingers for counting. The ordinary people, along with the merchants and traders, of almost every part of the world have used finger-counting at one time or another. In some places it is now largely a matter of historical interest, but in others it is still a significant aspect of contemporary culture.

Counting by Twos

Counting is theoretically possible using an entirely unsystematized sequence of unrelated number-words. The tax on the memory is, however, very quickly unacceptable. There are still a few undeveloped peoples who have no system of number-words. The most primitive of these have been found to have specific words only for one and two; anything above two becomes *many*. A frequently quoted example is that of the Damara of Namibia, who were prepared to exchange more than once a sheep for two rolls of tobacco but would not simultaneously exchange two sheep for four rolls.

Here, it is important to sound a certain note of warning. We must not jump to the conclusion, as some anthropologists have done, that people have no appreciation of numbers beyond the limit of their number-words. Words are very often found to be accompanied by gestures. Both must be properly investigated before general conclusions are drawn. Often other factors play a part in number awareness, and these must also be taken into account.

We can still find evidence amongst undeveloped peoples of a restricted basic sequence of number-words comprising words for one and two only but extended by the combination of these as far as six or eight. Such evidence has been found in Africa, Australia and South America.

The Bushmen of Botswana are perhaps the oldest people in all Africa.

Their spoken number-words proceed:

one:	*a*	four:	*oa–oa*
two:	*oa*	five:	*oa–oa–a*
three:	*ua*	six:	*oa–oa–oa*

Beyond six, they use a word meaning many. A similar kind of sequence has been found spoken by a tribe of aboriginals living near the Torres Strait between Australia and New Guinea. This goes:

one:	*urapon*	four:	*okosa–okosa*
two:	*okosa*	five:	*okosa–okosa–urapon*
three:	*okosa–urapon*		

An example from South America is:

one:	*teyo*	four:	*cayapa–ria*
two:	*cayapa*	five:	*ciajente*
three:	*cho–teyo*		

It we look at these examples, we are immediately struck with the way in which the word for two plays a special role. Each sequence builds up, with perhaps some elisions, in the form

one	two–one
two	two–two (or two-again)

In the first two cases five is expressed by the form 'two-two-one'; in the third case we have the word *ciajente* for hand. The African example has six expressed by the form 'two–two–two'. Beyond the words quoted, there is in each case a word meaning many. All three cases, however, are examples of two as a base for counting.

A number of reasons have been put forward for the existence of systems of counting by twos; the most obvious lies in the fact that man has two hands, two feet, two eyes, two ears, and so on. Another suggestion is that two-counting systems originate from ancient fertility rituals in which each male has to pair with a female. There is support for this in that the number-words of an Old Sumerian sequence actually mean:

man
woman
many

Less plausible are suggestions that they originate in man's early awareness of in front and behind – what he can see and what he cannot see, or in his psychological appreciation of the I–thou situation. Whatever their true origin, however, there is strong evidence that such systems were once widespread.

In theory, the two-counting principle is extendible as far as is needed. We just follow the pattern:

one	two–two–one
two	two–two–two
two–one	two–two–two–one
two–two	

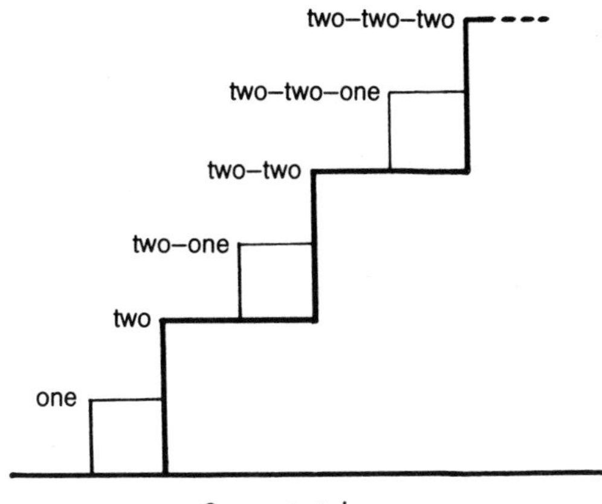

2-count staircase

and so on. But if we actually try this out in practice, it is not long before we have difficulty in remembering how many times *two* has been repeated. So it is not surprising that those examples which have been found go only as far as five or six, or occasionally as far as eight. Without an independent system for counting the twos there are insuperable difficulties.

Counting by twos can, however, be termed a number-word system, even if only a rudimentary one. A system can be said to exist once there is a basic sequence of number-words together with some kind of systematic way of combining them to extend this sequence. As soon as independent number-words for one and two are available, a system becomes possible.

We can represent counting by twos pictorially by means of a staircase with two kinds of steps, large steps or landings for twos, and small steps for ones. To reach five, we first take two large steps and then follow these with one small step. This represents precisely what is done when a number-word of the form 'two–two–one' is spoken.

If we assume that these examples of two-count correspond to the earliest counting system, it is reasonable to look at other systems found amongst undeveloped tribes for the next stage of development. Clearly, the problem of incessant repetition of the word for two can be avoided by having more than two independent number words.

One of the aboriginal dialects of Australia has the following number sequence:

one:	*mal*	four:	*bularr–bularr*
two:	*bularr*	five:	*bularr–guliba*
three:	*guliba*	six:	*guliba–guliba*

Here we see independent number-words as far as three, followed by the combinations:

two–two
two–three
three–three

The Toba tribesmen of Paraguay count a little differently:

one:	*nathedac*	six:	*cacayni–cacaynilia*
two:	*cacayni* or *nivoca*	seven:	*nathedac–cacayni–cacaynilia*
three:	*cacaynilia*	eight:	*nivoca–nalotapegat*
four:	*nalotapegat*	nine:	*nivoca–nalotapegat–nathedac*
five:	*nivoca–cacaynilia*	ten:	*cacayni–nivoca–nalotapegat*

They have effectively independent number-words as far as four, though *cacaynilia* is clearly derived from a two–one form. After four, the sequence follows the pattern:

two–three	two–fours
two–threes	two–fours–(and)–one
one–(and)–two–threes	two–(and)–two–fours

There is an interesting innovation here, the appearance of the multiplicative principle. 'Two–three' is additive – it means two plus three; 'two–threes' is multiplicative – it means two times three. *Cacayni–cacaynilia* (six) and *nivoca–nalotapegat* (eight) are also clearly multiplicative forms.

There are many variations in the way such systems are built up. The most important point is that they all share evidence of counting by twos. However, they are clearly not examples of pure two-counting and we need some general term under which they can all be collected. The term often used is neo-two-counting. It can be found in Africa, the Americas, southern Asia, Australia and New Guinea, usually in places which are close to pure two-counting, where the latter still exists.

There is an interesting variation in neo-two-counting systems in which odd numbers are expressed by subtraction. There are several examples of seven being expressed as 'one–(less-than)–two–fours' and nine in the form 'one–(less-than)–two–fives'. The reference point in forming these number-words is thus to the next pair rather than to the previous one. It is equivalent, as it were, to mounting the larger steps of the two-counting staircase until the required number has just been exceeded and then taking a single smaller step down. This is, however, a comparatively rare phenomenon in neo-two systems of counting. It is less rare in systems where counting by pairs has become so closely associated with finger-counting that the forms 'two–two–one' of pure two-counting and 'two fives' of neo-two-counting have been replaced by words meaning hand and two hands.

The reason for the introduction of the subtractive principle is not known, though a number of theories have been suggested. It is more easily explained where five and ten are associated with one and two hands. Here, it is not unreasonable to suggest that two hands had a special significance and that it was therefore natural to think of nine as being almost two hands. Indeed, there is additional support for this in other examples where eight, seven and six are expressed in the forms 'two–(less-than)–ten', 'three–(less-than)–ten' and 'four–(less-than)–ten'. This, however, does not explain why seven should be expressed in the form 'one–(less-than)–two–fours' rather than 'two–threes–(and)–one' in neo-two systems. The only plausible explanation is that at one time in such systems eight was a limit of counting above which a word for many was used and for this reason was of special significance.

The distinctive feature of neo-two counting is the introduction of pairing numbers together to provide forms such as 'two–threes' or 'three–three' for six instead of the pure two-counting form 'two–two–two', following the pattern of 'two–two' for four. We have also seen that this form of pairing occurs with certain finger-counting systems. It seems not unreasonable to suggest, therefore, that neo-two-counting developed as a variation from pure two counting as a direct result of the introduction of counting on the fingers. This receives further confirmation from examples where five is directly expressed by words meaning hand.

Counting by Fives

With the arrival of finger-counting, and also counting on hands and feet, the way became open for the abandonment of two as a base for counting and its replacement by five. This provides a further way of avoiding the repetitions of two and neo-two systems. We have already noted the replacement of the pure two-counting form 'two–two–one' by a form meaning hand. Finger-counting immediately suggests the use also of 'two–hands', 'three–hands', and so on, when counting ten, fifteen, and further multiples of five objects. Use of the feet as well as the hands suggests the forms 'two–hands–and–one–foot', 'two–hands–and–two–feet' for fifteen and twenty. There are many well developed examples of such systems to be found in evidence collected over the last hundred years or so.

Consider the following sequence of number-words to be found amongst South American Indians:

one:	*tey*	nine:	*teyente–cajezea*
two:	*cayapa*	ten:	*caya–ente*
three:	*toazumba*	eleven:	*caya–ente–tey*
four:	*cajezea*	...	... ...
five:	*teente*	fifteen:	*toazumba–ente*
six:	*teyentetey*	sixteen:	*toazumba–ente–tey*
seven:	*teyente–cayapa*	...	... ...
eight:	*teyente–toazumba*	twenty:	*cajezea–ente*

The form of the words is quite clear; they follow the pattern:

one	hand–two	three–hands
two	hand–three	three–hands–one
three	hand–four	... ...
four	two–hands	four–hands
hand	two–hands–one	
hand–one	... ...	

There is no doubt that this particular system is due to counting on the hands only.

Another example from South America includes the following number-words:

...	...	...
five:	*amgnaitone* (whole–hand)	
six:	*itacono–amgna–pona–tevinitpe* (one–on–the–other–hand)	
...	...	...
ten:	*amgna–aceponare* (two–whole–hands)	
eleven:	*puitta–pona–tevinitpe* (one–on–the–foot)	
...	...	...
sixteen:	*itacono–puitta–pona–tevinitpe* (one–on–the–other–foot)	
...	...	...
twenty:	*tevin–itoto* (one–man)	

This system is clearly derived from counting on hands and feet. It is particularly interesting because it goes on to express twenty–one in the form 'one-on-the-hand-of-another-man'.

Such examples of counting by fives have been found amongst indigenous tribes in most continents, but especially in Africa, certain parts of Asia, and the Americas. They derive from a stage of development when man is counting quite effectively up to twenty and beyond but there is still no evidence of any abstract appreciation of number. The number-words themselves, except for the first few when these are not related to the names of fingers, are just describing counting-gestures involving the hands·or the hands and feet.

As with two-counting, it is theoretically possible to extend counting by fives as far as is needed. The pure five-counting staircase follows the pattern of that

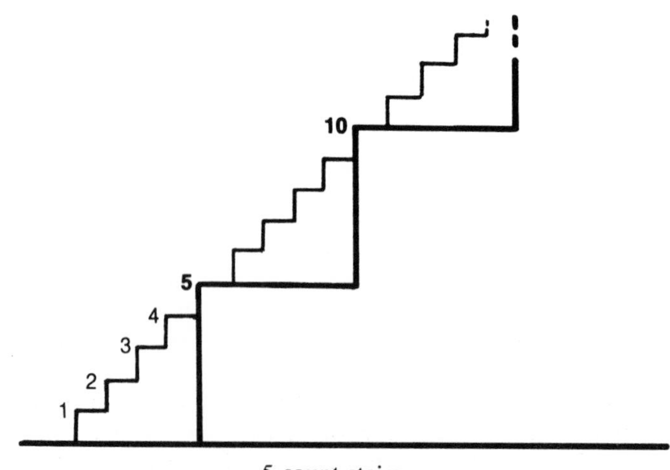

5-count staircase

for two-counting except that the large steps or landings have four small unit steps between them. However, no example of a number-word system has so far been found where pure five-counting is extended to large numbers. There is an obvious reason for this, namely that larger natural groupings of ten or twenty automatically suggest themselves once a system of counting by fives is developed. We can see an example in the last number-words quoted, where the word for twenty means one man and not four hands or two hands and two feet. Grouping in tens or twenties is found wherever five-counting is well established.

Initially, the systems of counting by fives and tens and counting by fives and twenties look much the same. They may well have a near-identical form as far as nine. We would expect, however, that there would be a clear distinction at ten, the five–twenty systems expressing ten as 'two–fives'. Unfortunately, this is by no means always the case. It is therefore safer to distinguish between these two systems by looking at the way in which multiples of ten from thirty onwards are expressed.

Pure forms of five–ten- and five–twenty-counting systems proceed by fives using words corresponding to the following forms:

five–ten-counting	*five–twenty-counting*
five	five
ten	two–fives
ten–and–five	three–fives
two–tens	twenty
two–tens–and–five	twenty–and–five
three–tens	twenty–and–two–fives
three–tens–and–five	twenty–and–three–fives
four–tens	two–twenties
four–tens–and–five	two–twenties–and–five
five–tens	two–twenties–and–two–fives

and so on in each case. There is no possible confusion between the two systems once ten is reached. However, a great many of what are basically five–twenty systems proceed in the form:

five	twenty–and–ten
ten	twenty–and–ten–and–five
ten–and–five	two–twenties
twenty	two–twenties–and–five
twenty–and–five	two–twenties–and–ten

and so on. This represents a basic five–twenty system with grouping at ten superimposed. In all three cases the intervening number-words are formed by adding words for one, two, three, and four to the appropriate five-group word.

The staircases representing the five–ten and five–twenty systems do not have the same overall form. For the former, we have large landings with one smaller landing between each successive pair. For the latter, we have large

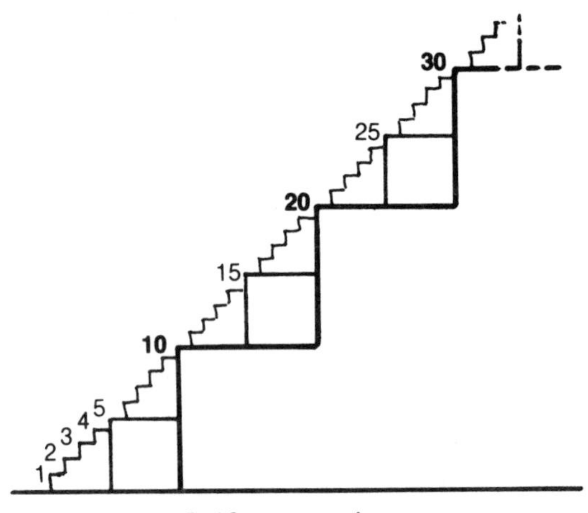

5–10-count staircase

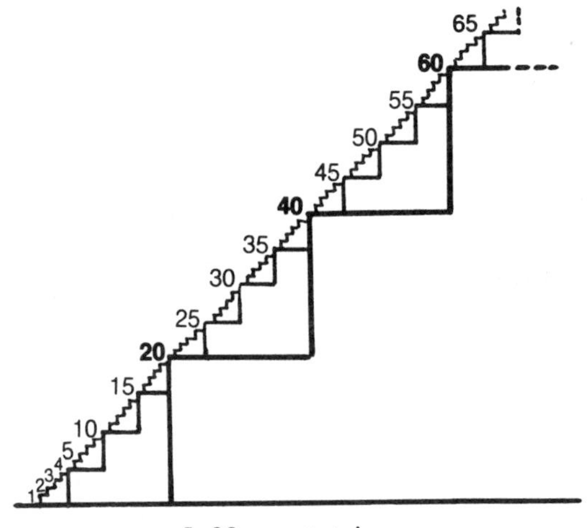

5–20-count staircase

landings with three smaller landings between each successive pair. Thus, in the five–ten system we get to the number-word for thirty-six by taking three large landing-steps followed by one smaller landing-step. In the five–twenty system, the unit step is taken after a single large landing-step followed by three smaller landing-steps. Occasionally we have examples of the subtractive principle, and there are many variations in the order in which individual number words are combined. Fifteen, for example, is found in the form 'five–and–ten', though not quite as often as in the form 'ten–and–five'.

There are many cases of developed five systems which are worth quoting, but we shall have to restrict ourselves to just one or two examples. A fairly typical five–twenty system is that of the Aztecs of Mexico. This goes:

one:	*ce*	...	... ...
two:	*ome*	fifteen:	*caxtulli*
three:	*yey*	sixteen:	*caxtulli–on–ce*
four:	*naui*	...	... ...
five:	*macuilli*	twenty:	*cem–poualli*
six:	*chica–ce*	...	... ...
seven:	*chic–ome*	thirty:	*cem–poualli–om–matlacti*
eight:	*chicu–ey*	...	... ...
nine:	*chic–naui*	forty:	*ome–poualli*
ten:	*matlacti*	...	... ...
eleven:	*matlacti–on–ce*	fifty:	*ome–poualli–om–matlacti*

and so on. *Cem–poualli* just means one twenty, *ome–poualli* two twenties. It is clearly a five–twenty system since thirty is spoken in the form 'one–twenty–plus–ten' and not in the form 'three–tens'.

By contrast, if we examine the counting system of the Luo of Kenya, we find the five–ten principle. Their sequence goes:

one:	*achiel*	...	... ...
two:	*ariyo*	ten:	*apar*
three:	*adek*	eleven:	*apar–achiel*
four:	*angwen*	...	... ...
five:	*abich*	twenty:	*piero–ariyo*
six:	*ab–chiel*	...	... ...
seven:	*ab–ariyo*	thirty:	*piero–adek*

and so on. This is a five–ten system. The clue lies in *piero-adek* (thirty); this is expressed as 'ten–threes' and not as 'twenty–and–ten'. The five-counting is evidenced by *ab–chiel*, *ab–ariyo*, and so on, taking the forms 'five–(and)–one', 'five–(and)–two', etc. This exactly parallels *chica–ce*, *chic–ome*, etc. of the Aztec five–twenty system.

Evidence of five–ten- and five–twenty-counting systems has been found in most continents of the world. There does seem to be something of a dilemma, however, when we care to try to decide which of these is the older. On the one hand there is a temptation to argue that the five–ten system is the older because man counted on his fingers only before counting on both fingers and toes. There is the opposite possibility, however. If we suppose that the five–twenty count was invented after finger–and–toe counting but before man began to make use of footwear, might not five–ten count be a later development when toe counting became difficult because of such footwear? This second argument runs into difficulty as we have to account for the fact that some Eskimo tribes, who presumably have never had access to their toes because of the cold, have a five–twenty system. In fact, neither argument is particularly helpful for the historian, being much too speculative. Fortunately, there is more convincing evidence to be had from the distribution of the various counting systems amongst the indigenous peoples of the world. Although the five–ten and five–twenty systems of counting survive today only amongst undeveloped peoples and are gradually disappearing as these peoples develop economically and have increasing contact with the so-called civilized world, the five–ten system found a new lease of life with the development of the abacus. Although basically decimal computers, many abaci have an intermediate collective bead or counter representing five units in addition to those denoting the usual powers of ten. This is a topic to which we shall return in Chapter 5.

Counting by Twenties

So far, our discussion of counting by twenties has taken account only of those systems where there is clear evidence of counting by fives. There is, however, much evidence available which points to twenty-counting without at the same time suggesting specifically a five–twenty system.

Many vestiges of twenty-counting can be found amongst the number-words of contemporary decimal systems. The most familiar to us is probably the English word *score* which, incidentally, provides a link with counting by cutting (scoring) notches on a tally stick. Another familiar example is the French *quatre-vingt* for eighty, and there are many others, especially amongst the Celtic languages. The Basque number sequence is very largely built up on the twenty-counting principle. A somewhat extreme example is provided by Danish number-words:

...	...	forty:	*fyrre*
three:	*tre*	fifty:	*halvtres*
four:	*fire*	sixty:	*tres*
five:	*fem*	seventy:	*halvfjers*
...	...	eighty:	*firs*
ten:	*ti*	ninety:	*halvfems*
twenty:	*tyve*	hundred:	*hundrede*
thirty:	*tredive*		

At first sight what catches our eye is the appearance of the prefix *halv* in the words for fifty, seventy, and ninety. In fact several of these number-words are abbreviated from longer forms. The full forms are:

forty:	*fyrre–tyve*	seventy:	*halv–fir–sinds–tyve*
fifty:	*halv–tred–sinds–tyve*	eighty:	*fir–sinds–tyve*
sixty:	*tre–sinds–tyve*	ninety:	*halv–fem–sinds–tyve*

Now the word *tyve* was originally just the plural of the word for ten, but it came to have the special meaning of two tens, that is, twenty. The Danish number words for thirty and forty retain that meaning of tens, but sixty and eighty are clearly represented by the forms 'three–times–twenty' and 'four–times–twenty'. Now *halv–tred*, *halv–fir*, *halv–fem*, which form the first parts of the words for fifty, seventy, and ninety, are to be understood as meaning a half less than three, four and five respectively. Thus *halv–tred–sinds–tyve* has the form 'two–and–a–half–times–twenty' and the words for seventy and ninety are to be interpreted similarly.

A clear example of twenty counting, also without evidence of counting by fives, comes from Nigeria. The Igbo number words include:

...	...	thirty:	*ohu–na–iri*
two:	*abuo*	forty:	*ohu–abuo*
three:	*ato*	fifty:	*ohu–abuo–na–iri*
four:	*ano*	sixty:	*ohu–ato*
five:	*iso*	seventy:	*ohu–ato–na–iri*
...	...	eighty:	*ohu–ano*
ten:	*iri*	ninety:	*ohu–ano–na–iri*
twenty:	*ohu*	hundred:	*ohu–iso*

The form of the sequence needs little comment; the multiplicative and additive forms are easily distinguished. The sequence continues by twenties, two-hundred being expressed as *ohu–iri*, 'twenty–(times)–ten'.

The Mayan number-word sequence of Yucatan is of special interest. This includes:

one:	*hun*	thirty-one:	*baluc–tu–kal*
two:	*ca*	...	...
three:	*ox*	thirty-five:	*holhu–cakal*
four:	*can*	thirty-six:	*uaclahu–tu–kal*
five:	*ho*	...	...
six:	*uac*	forty:	*cakal*
seven:	*uuc*	forty-one:	*hun–tu–y–oxkal*
eight:	*uaxac*	...	...
nine:	*bolon*	fifty:	*lahu–y–oxkal*
ten:	*lahun*	sixty:	*oxkal*
eleven:	*buluc*	seventy:	*lahu–cankal*
twelve:	*lah–ca*	eighty:	*cankal*
thirteen:	*ox–lahun*	ninety:	*lahu–y–hokal*
fourteen:	*can–lahun*	hundred:	*hokal*
fifteen:	*ho–lahun*	...	...
...	...	hundred-and-twenty:	*uackal*
twenty:	*hunkal*	...	...
twenty-one:	*hun–tu–kal*	two-hundred:	*lahun–kal*
...	...	three-hundred	*ho–lhu–kal*
thirty:	*lahu-cakal*	four-hundred:	*hun–bak*

This sequence deserves study. If we look at the number words up to *hunkal* (twenty), we find that they follow a roughly similar pattern to our own. There are independent number words as far as eleven, followed by the forms:

ten–(and)–two	four–(and)–ten
three–(and)–ten	five–(and)–ten

and so on. It would seem that this is a decimal counting system. However, *hunkal* has the form 'one–(times)–twenty', and this is followed by *hun-tu-kal* which has the form 'one–on–twenty'. The remaining number-words have the forms:

thirty:	ten–(towards)–two–(times)–twenty
thirty-one:	eleven–on–twenty
...	...
thirty-five:	fifteen–(towards)–two–(times)–twenty
thirty-six:	sixteen–on–twenty
...	...
forty:	two–(times)–twenty
forty-one:	one–towards–three–(times)–twenty
...	...
fifty:	ten–towards–three–(times)–twenty
sixty:	three–(times)–twenty
seventy:	ten–(towards)–four–(times)–twenty
eighty:	four–(times)–twenty
ninety:	ten–towards–five–(times)–twenty
hundred:	five–(times)–twenty
...	...
hundred-and-twenty:	six–(times)–twenty
...	...
two-hundred	ten–(times)–twenty
three-hundred:	fifteen–(times)–twenty
four-hundred:	one–(times)–four–hundred

It is easy to see that this is basically a twenty-system with the independent words *kal* for twenty and *bak* for four–hundred (twenty squared). Indeed, there are the further independent words *pik* for eight–thousand (twenty cubed) and *calab* for one–hundred–and–sixty–thousand (the fourth power of twenty). The words for twenty–one, thirty–one and thirty–six just build appropriately on *kal* as we might expect, but those for thirty and thirty–five relate to *cakal* (forty). We might suspect the subtractive principle to be at work here, but this is not the case with *holhu-cakal* (thirty-five), which has the bare form 'fifteen–forty' (or, strictly, 'fifteen–two–twenty'). The words for thirty and thirty–five only make sense as 'ten–(towards)–two–(times)–twenty' and 'fifteen–(towards)–two–(times)–twenty'. Thus, whilst we still have the idea of ten and fifteen unit steps being taken after reaching the first twenty landing, these number-words explicitly refer to the next twenty landing, forty, towards which they are climbing. We soon find this interpretation confirmed; above *cakal* (forty) the words invariably refer to the next twenty landing. Thus the word for forty-one is *hun–tu–y–oxkal*, 'one–towards–three–(times)–twenty', and not *hun–tu–cakal*, 'one–on–forty'.

There are several possible explanations for the somewhat strange Mayan number-word sequence. It seems unlikely that it represents the superimposition of a twenty-system on an already existing decimal one. The most plausible explanation is that there was at one time a limited sequence associated with finger-counting represented by the independent number-words up to *buluc* (eleven), and that this was extended by the additive principle as far as nineteen. The twenty unit, associated with counting on hands and feet, was then introduced and continued as far as *cakal* (forty), which may have represented a limit beyond which there was no need to proceed for ordinary day-to-day affairs. The extension beyond forty up to the large numbers which could be formed using *pik* (eight–thousand) and *calab* (one–hundred–and–sixty–thousand) was an artificial sequence invented by the priests which eventually influenced the way in which thirty and thirty–five were expressed. Unlike the Mayan written numerals (which we discuss in Chapter 3) this artificial extension of the number sequence does not follow the pattern of the Mayan calendar where the twenty-system is broken in one place by a grouping of eighteen. We therefore conclude that it was invented for some religious purpose.

The three examples of twenty-counting which we have discussed all include evidence suggesting ten-counting. Indeed, no completely pure twenty-system is known. Such a system would have independent number-words up to twenty, making it possible to count as far as three-hundred-and-ninety-nine. New words would be required for each power of twenty from twenty squared onwards.

It is, of course, possible that a system of pure twenty-counting existed at some time or other, but it seems unlikely. The existence of independent number-words up to five and ten, and even to eleven (using a word meaning beyond), can easily be explained by reference to the names of fingers. Toes do not all have individual names, and this suggests a possible explanation for the absence of pure twenty-systems. However, we must remember that number-words do not necessarily bear any relation to names of fingers or other parts of the body.

We cannot doubt that systems of twenty-counting in some form or another were once widespread in regions which today have only decimal counting. The evidence is too extensive to be disregarded. What is perhaps surprising is the extent to which remnants of twenty-counting can be found in our familiar decimal systems once we begin looking for them.

The History of Counting

When we come to attempt a reconstruction of the history of counting from its earliest beginnings, the lack of primary source material for the greater part of the time involved presents us with great difficulties. We can, however, supplement the evidence provided by contemporary spoken words and by written number words with limited evidence from number symbols. Having identified the various counting systems from this evidence, we can then view them from two different points of view, one primarily geographical, the other purely linguistic. The latter approach involves examining the number-words of each language to see if we can find traces of words or word-forms from earlier counting systems as we do, for example, in the case of French *quatre-vingt*. The geographical approach involves studying the distribution throughout the world of the various different systems amongst indigenous peoples.

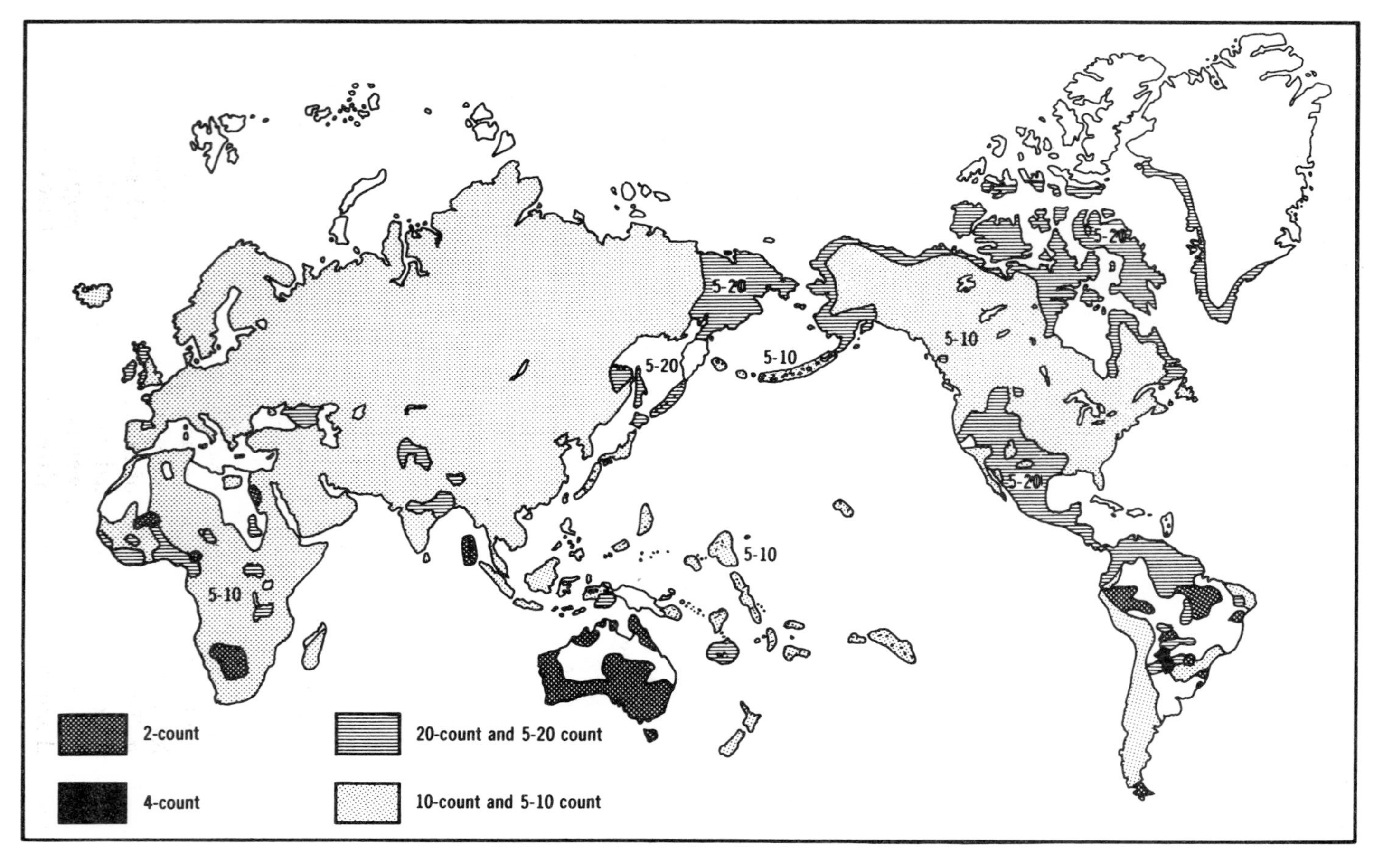

World map of counting systems

If we study a world map of the main indigenous counting systems, there are several things which immediately strike us; for example, we see that two-counting systems are found only on the edges of the continents. If we look at the two-counting regions in South America we see that they have twenty-counting regions to the North. Most of this twenty-counting is in the form of five–twenty systems, as an examination of the actual number-words will show. These five–twenty number-words include many examples where three is expressed in the form 'two–one' and four in the form 'two–two'. We can therefore infer that they have supplanted an earlier two system. The same situation can be found in Australia and to a more limited extent in Africa.

It is not unreasonable to conclude that man first of all became aware of the pairing principle and constructed his earliest counting system using the words for one and two only. This system of two-counting was probably invented in a few centres of civilization, or perhaps in one centre only, and then gradually spread over the world. With the advent of finger-counting it became possible to have a convenient grouping of units and so extend the two-system to meet the growing needs of society. Once invented, counting by fives rapidly superseded two-count almost everywhere and eventually developed into five–twenty-count, thus enabling larger numbers of objects to be counted.

It is tempting to see the five–ten system as a system of counting in its own right. A study of the distribution of five–ten systems on the map suggests, however, that it is a cross between five–twenty counting and the pure decimal system, the last system to be invented. We infer this because it occurs for the most part in regions adjoining both five–twenty and decimal counting. This inference is supported by a detailed study of the number-words of these two kinds of five-counting. Many five–twenty systems have the numbers six to nine formed by addition from five; few five–ten systems are complete in this respect. If our inference that five–ten systems are hybrid is correct, this otherwise puzzling phenomenon is easily explained. With the arrival of decimal counting, people speaking five–twenty number-words would first of all replace the twenty-system words for multiples of ten.

Even if we restrict ourselves to evidence obtained from indigenous peoples, we find that decimal counting is the most widespread system. Once it was invented, it must have spread comparatively rapidly so that other systems were very largely, though not entirely, replaced. Its success is not difficult to explain; it is the most convenient system so far invented and bears a direct relation to those parts of the body which are usually the most readily accessible, the hands. The older and less convenient systems are found today only amongst undeveloped peoples largely isolated from modern civilization.

If we assume that this somewhat sketchy history of the main-stream of development of counting is valid, we are still faced with the problem of accounting for a few rather specialized systems which do not appear to fit in easily with it. There are a number of examples of words which suggest that twelve-counting may once have existed. The English word *dozen* immediately springs to mind. There are also the well-known systems of geographical and astronomical units which are clearly sixty systems and which have come down to us from ancient Babylon. All such systems are later in origin than decimal counting, however, and are almost invariably associated with units of measurement, where divisibility is an important factor. They were thus developed primarily for purposes of calculation.

We are still faced with the problem of putting some rough date on man's becoming aware of numbers in the abstract, or at least suggesting where this might fit into our history of counting. Obviously, this would have been a gradual process which would have come about at different times in different places. In the case of some of the more undeveloped tribes to be found today, it has still not occurred. It is certainly a later occurrence than the invention of any of the counting systems which we have considered. Apart from evidence of tallying, written mathematical records go back approximately to 3000 B.C. Such ancient records are very largely associated with practical applications of number, though there are occasions where we find calculation carried out apparently for its own sake. It is reasonable to suggest that awareness of number in the abstract had its beginnings around this time or fairly soon after, though it did not become an influential part of mathematical thinking until the time of the Pythagoreans, that is, roughly the middle of the first millennium before Christ.

The Age of our Decimal Counting

We have not gone into the form of our own decimal counting system as we have assumed that this is too familiar to need detailed discussion. However, the question of its age is an interesting one and to answer it we have again to turn to linguistics. When we compare the various number words spoken in present-day Indo–European languages, we find that for the most part they can be traced back to common roots even though at first sight they appear very different. For example, the English words *five*, *ten* and *hundred* are easily traced back to corresponding Gothic words via Anglo–Saxon and Old Saxon. Gothic is the oldest of the Germanic languages. It was spoken by the tribes who originally lived near the Vistula but who eventually overran the countries of southern Europe. It survives in Bishop Wulfila's translation of the Bible, made in the middle of the fourth century during the Gothic occupation of Italy and now in the University Library of Uppsala. The Gothic words for five, ten and hundred are *fimf*, *taihun* and *hund*. Words in other Germanic languages can be traced back similarly, such as *vijf* (Dutch), *fem* (Danish and Swedish), *fünf* (German), all going back to Gothic *fimf*, and similarly with *tien* and *honderd* (Dutch), and so on. This enables us to identify a whole family of languages owing their origin to Gothic. This Germanic family includes English, Dutch, modern German, and the Scandinavian languages.

Another family, which we can also easily identify from number words, is the Celtic group. This includes Irish and Scottish Gaelic, Welsh, Cornish and Breton. Together with Gallic, the mainland Celtic language once spoken in Gaul, we can trace them back to a general Old Celtic stem. There is also the family of Romance languages: Italian, French, Spanish, Portuguese, Romanian and associated dialects. Then there is Greek, the oldest documented European language.

It is interesting for us to compare corresponding number-words from these different groups. We will look at words for five, ten and hundred:

Gothic	Celtic	Latin	Greek
fimf	*pemp*	*quinque*	*pente*
taih̥un	*dek*	*decem*	*deka*
h̥und	*kant*	*centum*	*hekaton*

Clearly, there are similarities, but there are also differences. The question therefore arises as to whether the different language families from which they are taken are or are not related.

Despite substantial differences in the words, a closer examination shows that there are rules of transposition by means of which we can establish a direct relationship. The connections between corresponding words in Celtic, Latin and Greek are clear; it is Gothic which seems to be the 'odd man out'. But if we apply the transpositions:

f ⟶ *p* or *q*
t ⟶ *d*
h ⟶ hard *c*

to the Gothic words, we find that we can establish a consistent relationship with the Celtic, Latin and Greek. This is the case also with other number words not considered here. It is from the fact that the transpositions are consistent and not haphazard that we deduce that these language families are related and so must go back to earlier roots. The overall family to which they all belong is known as the Kentum family, so called from the root word for hundred. Two Asian languages also belong to this family, Tocharian (once spoken in parts of Turkestan) and Hittite (once spoken in Anatolia).

We find another basic family of languages represented in Europe by Albanian and the Slavic and Baltic languages (with the exception of Finnish, which is related only to Hungarian). Sanskrit, Iranian, Armenian and other Near-Eastern languages belong to this family also. As before, we will look at examples of the words for five, ten, and hundred:

Russian	Lithuanian	Sanskrit
pjatj	*penki*	*panča*
dĕsatj	*dešimt*	*daśa*
sto	*šimtas*	*śatam*

Here again, we can discover consistent transformation rules which show conclusively that only one basic language family is involved. This is known as the Satem family, again from its root word for hundred.

When we compare the number words from the Kentum and Satem families, we notice immediately a consistent change from *k* or hard *c* to *s* or soft *c*. The consistency of this and certain other linguistic transformations tells us that these two families must themselves have been derived from a common root language or group of dialects. This ultimate root language has not been preserved in any form directly; we have to reconstruct it artificially. Since it has given birth to most, though not quite all, of the languages spoken from the Indian sub-continent in the East to the western islands of Europe, we call it Original Indo–European. Its origins are lost in pre-history, but it is clear that it had a decimal counting system from which many of our number words are descended.

We are able to estimate an approximate period for the origin of Original Indo–European by taking into account the large-scale movements of peoples within the areas in which Indo–European languages are now spoken and extrapolating backwards. This suggests that it dates from at least the fourth millenium before Christ. Our decimal counting system must date from the same era. This does not mean, however, that our Original Indo–European ancestors developed the whole decimal system which we use today. It seems certain that their largest independent number word was the ancestor of Greek *chilioi* and Latin *mille* (thousand), and that, if there was a need to express larger numbers than this, it was done by compounding the decimal number words for numbers up to thousand.

Although we have been able to trace *five*, *ten* and *hundred* directly back to Original Indo–European, there is no way in which we can relate *thousand* to *chilioi* and *mille*. *Thousand* has its origin in the Gothic word *pusundi*, which means literally strong hundred. *Million* derives from the Italian word *millione*, invented in the fourteenth century. *Billion*, *trillion*, and so on, appear in a manuscript of the French mathematician, Nicolas Chuquet, dating from 1484, though some were invented some years earlier. The Ancient Greeks had *myrioi* for ten-thousand derived from an earlier form meaning countless. In India, we find words in use as early as the first century before Christ for unbelievably high powers of ten. These were coined by the author of the *Lalitavistara*, but were for fanciful rather than practical use. The claim that our decimal counting system goes back to Original Indo–European ancestors must therefore be a qualified one. Certainly we can trace the decimal principle of our system as far back as we have suggested, and many of the number-words themselves up to *hundred* and, along with it, *thousand*. The extension of the system to incorporate independent words for ten-thousand and above is more recent.

It is interesting to note that even in the Indo–European decimal counting systems there is a hint of one–two–many counting in that early words for three can be linked to a root word whose meaning was beyond. If this evidence is correct, it would suggest that the Original Indo–Europeans themselves passed through the earliest stages of counting after the invention of the Original Indo–European language.

Counting with Knots

So far, we have been concerned with number-words and with counting on the hands, feet and other parts of the body. We shall begin the next chapter with an account of carving tally-marks, the first stage in the evolution of written numerals. The use of knots lies in between the use of parts of the body for counting and the carving of notches to make permanent records. It is something of a dead end in the history of numbers as it has never evolved into a spoken or written counting system. It has, however, been a fairly widespread practice.

The most developed system of knot numbers of which we have direct evidence is that of the Incas of Peru, whose empire flourished from the twelfth to the sixteenth century. Knots were tied on twisted woollen cords, often of different colours, as a means of keeping and communicating official state records as well as for ordinary commercial and domestic purposes. There was no other system for recording numbers. Some of these knotted cords, known as *quipus*, have been preserved and we can interpret the numbers indicated on them. Three different kinds of knots were used: ordinary single knots, figure-of-eight knots, and slip knots with from two to nine loops. Units were represented by a figure-of-eight knot for one and slip knots for two to nine. Tens and hundreds were represented by the appropriate number of single knots. Two-hundred-and-thirty-five would be represented on a single cord with two knots for the hundreds near the top of the cord, three knots for the tens lower down, and a slip knot with five loops at the bottom. A different cord would be used for each number represented; thus a number of sheep might be recorded on a cord of one colour, a number of goats on a cord of another colour, and so on. These different cords would each have a loop at the top by means of which they were all threaded onto a single head-cord. Often the head-cord would have knots representing the sum of the numbers on the cords hanging from it.

These Inca knot numbers have survived amongst the Indians of Peru and Bolivia, who thread fruit seeds onto strings in groups to represent decimal numbers. These *chimpu*, as they are called, have one very special characteristic: the power of ten is indicated by the number of threads passing through the appropriate number of seeds. Two-hundred-and-thirty-five requires three strings. Two seeds are threaded onto all three of these, three are threaded onto two, and five are threaded onto one only. The strings are then knotted together at both ends so that the number recorded can be preserved.

Knotted cords for recording numbers can be found in use amongst the North American Indians also. Sometimes these are used to denote the number of days until an important religious or social ceremony. In this case, simple knots only are used and there are as many knots on the cord as there are days to pass before the ceremony. A knot is untied or cut off each day and, when the last knot is reached, participants know that the important day has arrived. We find knotted cords used in this fashion in Africa also. A man going on a journey will give his wife a cord on which the number of knots indicates the days until he is to return. Pregnant women tie knots to remind them of the number of months before a baby is due; in such cases one knot will be removed at each full moon. Bookkeeping records are also kept in many parts of Africa with the aid of knots, though often the more formal transactions will be recorded by cutting notches on tally sticks leaving the purely domestic records to be kept with knot numbers. In a few regions, there is a convention whereby the women use knots and the men tally sticks.

We have evidence of knot numbers from the Far East, including the Indian sub-continent. The fifth-century Chinese philosopher, Lao-Tse, recommended the knot method for representing numbers. Chinese tablets have been preserved on which knots on white cords are depicted being used for odd numbers and knots on black cords for even numbers. Knots tied in straw are still used in islands off the Japanese mainland for recording time worked and wages due. In Tibet, knotted strings are occasionally used to count religious exercises, though more usually this is done by using a string of beads. It is possible that the way in which certain eighteenth-century Indian scripts were written is derived from counting by knots.

We can find evidence in Europe from both ancient and modern times. There are a number of references to knots in classical literature. One of the best known is to be found in Herodotus, who refers to a cord having sixty knots representing the number of days for which the Persian King, Darius, expected to be away on a journey. The knotted ropes used from the sixteenth century as part of the apparatus for measuring the speed of ships at sea has bequeathed to us the familiar knot, the nautical unit of speed. In Germany, Austria and some parts of Switzerland, knots were in use until the early part of this century for recording millers' transactions with the bakers to whom they supplied flour and meal. The Tibetan prayer strings have their counterpart in the knotted woollen cords used by Orthodox monks when reciting the Jesus Prayer just as the Tibetan threaded beads correspond to the Rosary of the Roman Catholic Church.

All these and other examples go to show how widespread has been the use of knots for numbers. However, in the specific context of counting, we should distinguish between those systems of knots which have been used solely for recording numbers and those used as an aid to counting. It is those used for recording only, often in association with the use of the abacus for computation, which, in a sense, belong to the undefined area between this chapter and the next. The quipu and the millers' knots are not strictly methods of counting; prayer cords, and wedding and pregnancy knots, are, though they are devised for specialized purposes. Often such counting is not associated with number words; the objective is simply to receive a reminder when the succession of knots runs out. Sometimes there is scarcely any intention to relate specifically to numbers. In such cases the knots differ little in principle from our custom of tying a knot in a handkerchief to remind ourselves of some important engagement.

Chapter Three
Writing down Numbers

> Our forefathers had no other books
> than the score and the tally.
>
> (Shakespeare)

> Let him that hath understanding
> count the number of the beast: for
> it is the number of a man, and his
> number is Six hundred three-score
> and six.
>
> (Apocalypse)

Scarcely a day passes in which we do not have either to read numbers or write them down. Written numbers are so familiar to us that we hardly give them a thought, except when we have to make a calculation or compare the displayed prices in shops. Even then, it is not the forms of the numbers (that is, the numerals themselves) which strike us, but the problem of manipulating them or the information which they convey. The symbols themselves we take for granted. They are international, and almost universal. Unlike number-words they transcend the barriers of language. This is, however, a comparatively modern phenomenon. Our numerals have evolved over a period of many centuries, during which they have travelled from India via the Arab world to Europe, replacing other forms of written numbers on the way. From Europe they have spread to all corners of the Earth, except for a few remote areas where modern civilization has not penetrated. But their triumph has not been total; other forms have survived along with traces of those which have disappeared. A study of the evidence enables us to map out a history no less complex than the history of number-words. In this case, however, we are much more fortunate with our evidence. The purpose of writing numbers is to make a lasting record, hence much of the evidence has survived even from early times. We are fortunate too in having a definite end-product in the familiar numerals which virtually all of us use today. This is in contrast to the situation with spoken numbers where differences of language mean many different sets of number-words and similarity is confined to the system of counting used.

Tallies

Most of us are familiar with tallying though we may not have thought seriously about the principle involved. It is unlikely that the word itself will be part of our everyday vocabulary, though other words associated with it almost certainly are. We can picture the bold bad man of the Wild West carving a notch on the handle of his gun each time he kills an opponent. Such a record of killings was not, however, exclusive to the land and times of Billy the Kid and the Lone Ranger. In the wars of the twentieth century it was not uncommon for a fighter pilot to paint an aircraft emblem on the side of his plane for each enemy shot down. In a similar way, bomber crews painted pictures of bombs on their machines, each bomb representing a mission over enemy territory successfully accomplished.

We can see the same principle in the two illustrations from primitive and hence supposedly much less civilized cultures. On the Fiji Islander's club, now in the Museum of Ethnology at Frankfurt-am-Main, we can see how notches have been carved to record the number of enemies laid low by its owner. Again, on the sword from the Philippines, now in the Linden Museum at Stuttgart, we can see the silver nails carefully inlaid in the blade for the same purpose.

Coming nearer home, we may recall the evenings spent at our favourite local inn. If we are regular and respected customers, it is likely that the

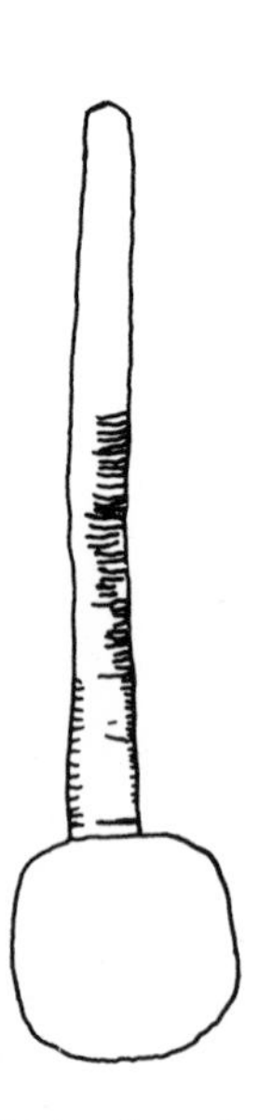

Fiji islanders club

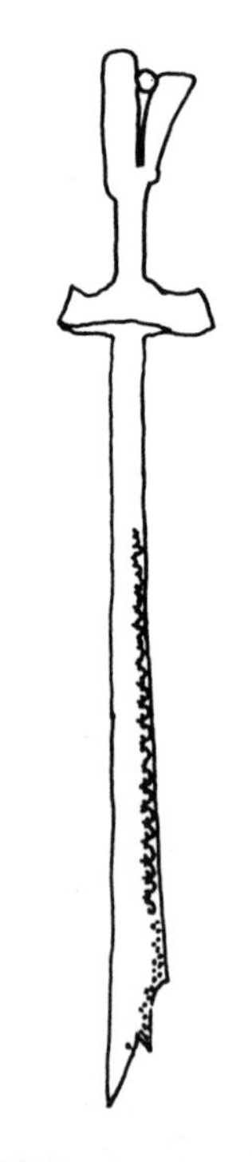

Philippine sword

barman will keep a tally of the pints which we consume instead of insisting that we pay for every drink when it is served. He may well note down on a slate or a card the pints which he draws for us by recording a simple stroke I on each occasion so that, after we have four pints, the record standing against us will look like IIII. In fact, he will be using the oldest principle for recording numbers known to man, namely the principle of the one-to-one correspondence of a series of identical marks with the objects which they are intended to represent. Each stroke corresponds to one pint of beer, just as each silver nail on the Philippine sword and each notch on the Fiji club or the bad man's gun corresponds to a victim despatched, each aircraft emblem to an enemy shot down, and each painted bomb to a mission successfully completed. This, of course, is almost the same as the principle of counting on the fingers, discussed in Chapter 2, the difference being the making of a lasting record.

The need for permanent records for numbers of things must have arisen very early in the history of mankind. Certainly, this need was earlier than the invention of any written number-words and possibly even earlier than the first number-words themselves. Making a tally by holding up fingers or by scratching marks in the earth or sand or by piling up heaps of pebbles leaves no permanent record, though each of these and similar methods does provide a simple visual presentation of numbers by means of one-to-one correspondence. As man passed from the stage of gathering and storing food to its actual production and from the hunting and killing of animals to the keeping of herds, the early and temporary tallying methods became inadequate for everyday purposes. Man therefore had to resort to carving notches on whatever was most handy and suitable, such as sticks of wood and bones of dead animals.

At some stage an attempt would have been made to store records of numbers by keeping tallies of heaps of sticks or stones in a safe place, perhaps in some kind of box or other container. We do know, for example, that one method of counting a body of soldiers which was actually used in ancient times was to march the men in single file through a gateway or past a stake driven into the ground and to put one stone in a box for each soldier that marched by. Presumably such boxes of stones could have been stored away so that records could be kept; but an official, opening such a box and looking into it, would get no immediate visual picture of the number which had been recorded. Unless that number was very small indeed, the stones would first have to be set out in a row or in some sort of pattern. An accidental upturning of a box of stones might well cast serious doubts on the accuracy of the record.

We know that tallying is of very great antiquity by the survival of evidence from both the Old and the New Stone Ages. One particularly interesting find, now in the Moravaské Museum in Czechoslovakia, is the palaeolithic wolf bone found in 1937 by Dr Karl Absolon during excavations in Vestonice (Moravia). The bone is about 18 cms long and has fifty-five notches in it. There are twenty-five notches, all roughly equal in length, followed by a single notch approximately twice the length of the others. Then there is a

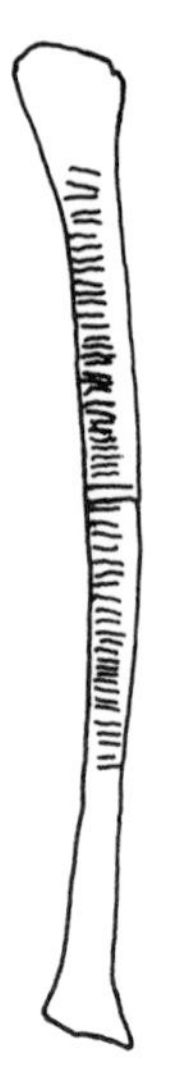

Wolf bone explanatory sketch

second long notch with a further series running as far as thirty. We shall never know just what these notches were intended to represent, though we can be pretty certain that it was numbers of some kind. Nevertheless, the wolf bone does give us clear evidence that the tallying principle for numbers goes back at least thirty-thousand years. Also, the grouping in fives suggests that even at that time there was a connection of some sort between the permanent recording of numbers and counting on the fingers. There is no way, however, in which we can come to a clear conclusion as to whether or not any corresponding number-sounds had been invented. We do know that some forms of art had been developed by the Diluvial peoples of those times because a sculptured ivory head of a woman, contemporary with the wolf bone, was excavated near it.

Another example, this time from Africa and dating from some nine thousand years before Christ, is the bone tool handle discovered by Dr Jean de Heinzelin at Ishango on Lake Edward in Zaire. It has a piece of quartz fixed in a narrow cavity at one end, the tool probably having been used for some kind of engraving. On the handle we can see three rows of notches arranged in groups of different numbers. One row has groups of eleven, thirteen, seventeen and nineteen notches, and another has groups of eleven, twenty-one, nineteen and nine. All sorts of speculations suggest themselves: for example, can we infer from the row of eleven, thirteen, seventeen and nineteen notches – these numbers being the prime numbers between ten and twenty – that there was some sort of development of elementary number theory in those far off times? Again, there is a third row which seems to suggest a kind of two-times multiplication table. A group of three notches is closely followed by a group of six, then a gap and four notches followed by

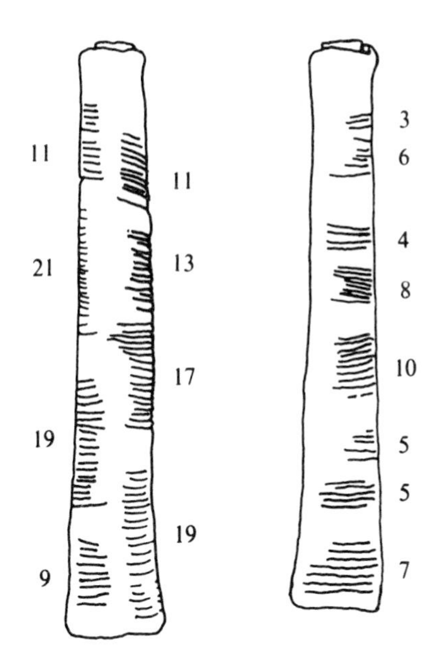

Ishango tool handle explanatory sketch

eight, then after a further gap, ten followed by five. This suggested interpretation encounters difficulties, however, because we now find a group of five notches followed by one of seven, though it is tempting for us to see these last two groups being continued by the row with groups of eleven, thirteen, etc. and providing re-inforcement for the prime number speculation. In fact, it is virtually impossible to discover what these various groups of numbers signify. There is almost certainly some kind of connection with finger-counting and also with the ways in which simple calculations were carried out. When we recall the examples which we quoted in Chapter 2, where even numbers such as six were expressed in the form 'two–(times)–three' rather than 'five–(and)–one', there is a strong temptation to associate this with some of the groupings on the bone.

It is always dangerous to jump to conclusions without corroborating evidence. Deductions about the extent to which peoples, so long separated in time from us, had developed their understanding of numbers cannot be made on the basis of the Ishango bone alone. However, from this find and many others like it from various parts of the world, we can deduce that the recording of numbers by notches carved on suitable objects is of very great antiquity and was virtually a universal practice. Its usefulness is demonstrated by its survival in various guises right up to the present day despite the availability for many centuries of much more sophisticated and efficient methods of writing numbers. We can also surmise that the method is much older than actual archaeological finds suggest, since the earliest tallyings carved on trees or pieces of wood could not have survived the intervening ages in the same way as records on more permanent materials such as bone or metal.

Of course, the barman at our local inn could just as easily record our drinks by keeping a running total using modern numerals:

1234 . . .

But these are a comparatively late invention and have only been the common possession of the ordinary working man, even in developed countries, for a short period of time. In the Middle Ages, although there was a well developed system of numerals available, it was known only to the privileged who had had the benefit of the education provided by the schools run by the religious institutions of the day. The ordinary peasant or labourer did not know how to read or write. The acquisition of such special skills was largely unnecessary for him as there was no printed material for him to read. The kind of material which was so laboriously copied out by the monks was the preserve of a small number of people, originally almost exclusively the nobility and the clergy, but later extending to the various merchants and traders as well as to lay scribes and keepers of public records. Nevertheless, the ability to keep some kind of written record of numbers was essential for the unprivileged and unlearned. After all, amounts of crops, numbers of cattle, totals of hours worked, and details of barters and financial transactions had to be kept in an age when the struggle for existence was often hard, taxes of various kinds had to be paid, and a proportion of working time had to be spent in the service of highly demanding though often absentee local overlords. For such people tallying was the obvious and the only method of keeping personal records. Only in this way could they provide some measure of protection against entirely arbitrary demands, often made by unscrupulous factors, which if sufficiently excessive could deprive a man and his family of the few possessions and what little security they had.

Of course, cheating by adding additional cuts to an existing tally stick was simple and must have been widespread. There was thus a need for the duplication of records so that both parties to a barter might have their own record of a transaction, especially if there was to be any question of accumulating debts. The identical marking of two different tally sticks did not prove adequate for the prevention of fraud, since a debtor could always be accused of simply replacing his tally stick with another on which fewer notches had been cut. Equally, it was a simple matter to cut in additional notches where this was advantageous.

At some time or other the idea was conceived of splitting a tally stick lengthways down the centre after the notches recording various amounts had been incised. Protection against alteration was then provided as the two

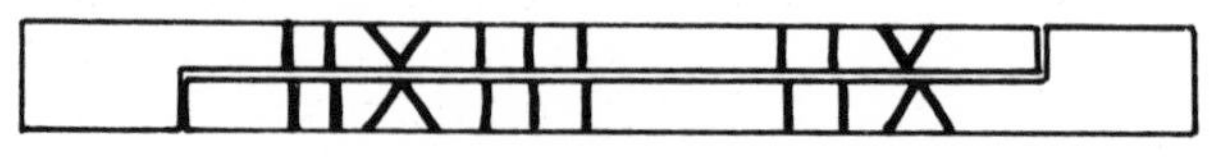

Split tally stick

pieces of wood could be carefully matched against each other and any discrepancy between them would immediately be revealed. We cannot say how old this practice of splitting tally sticks is. All the split sticks which have survived are of comparatively recent origin, and so, whilst they provide evidence of the survival of the practice, they do not provide evidence of its origin. Familiarity with the nature of man would suggest that it may be of considerable antiquity!

We may be surprised to learn that tally sticks have been used until quite recent times by government departments for the keeping of records of taxation payments. During repairs to Westminster Abbey in 1909 a number of surviving British Exchequer tallies were unearthed. These are of the split kind, and have notches of different sizes carved on them representing different denominations of currency. Officially, their use was forbidden by statute in 1783, but they were not entirely replaced by handwritten records and the use of paper money until about the end of the first quarter of the nineteenth century. Most of the then surviving exchequer tally sticks were put to supposedly good use feeding the furnaces heating the House of Lords. Unfortunately this led to a general fire and the destruction of the old Houses of Parliament. The discovery of the tally sticks in Westminster Abbey was particularly fortunate for the historian. There may well be others still in existence which have so far not come to light.

We can find direct evidence of the use of tally sticks in many European countries and indirect evidence in almost all of them. They were used in Russia as well as in Britain for the recording of taxes. In some countries, such as Switzerland, they remained in use for agricultural purposes, such as registering grazing rights, until well into this present century. Many of these tally sticks have survived and are to be found in national and local museums. There are numerous references in European literature and in town and city statutes. Most of these confirm the widespread use of tally sticks for registering debts and other financial agreements between individuals as well as for official state bookkeeping. We noted in Chapter 2 that the English word *score* provides a direct link with tallying in that it also means a cut. Many European words mean both a debt and a cut, and many phrases expressing the owing of goods or money make use of words which once meant cutting a tally mark, and in some cases still retain this meaning. We find the same with the classical languages. In Latin the word *putare* meaning 'to cut' led to both *imputare*, which took on the meaning 'to assign a debt', and *computare*, 'to calculate'.

Tallying is still important in many other cultures, usually for recording business transactions. Sometimes, we find quite unique applications. Men of the Songe tribe of Zaire, for example, record each journey which they undertake by carving notches on their walking sticks. In several areas the yearly calendar was recorded by painting or scratching the days and lunar months on the mud walls of homesteads, one stroke for each day and a larger stroke for the months. It has been recorded that in Buganda, tallies once took the form of boards with rows of holes. Small holes into which white pegs were placed denoted units, larger holes with black pegs tens, and still larger holes with red pegs hundreds. If we turn to the Far East, we again find that the evidence is

Chinese character for 'contract'

plentiful. In some cases, tally sticks are still in use. Elsewhere we find the familiar indirect evidence such as provided by the Chinese character for 'contract' which includes characters denoting 'notched stick' and 'knife'.

It is unfortunate for us that wooden tally sticks are so easily lost or destroyed. We can be fairly sure that wooden sticks were used for the recording of numbers of objects long before the time of the bone tallies which have survived and which themselves date from prehistoric times. Cutting notches on wooden sticks is thus almost certainly the first stage in the history of the recording of numbers in a permanent or semi-permanent form. Much of the earlier history of tallying has, however, to be reconstructed from evidence which is very scarce indeed and which cannot be truly representative of the different kinds of materials used. Nevertheless, the examples which we have are important for the history of number because tallying represents the first beginnings of the invention of distinctive symbols for numbers as well as providing direct evidence of the great antiquity of the use of the principle of one-to-one correspondence.

The Dawn of Numerals

We can see just how widespread is the evidence for the evolution of the first number symbols from notches on tally sticks by looking in detail at some of the most ancient systems of numerals of which we have direct historical evidence. In almost all of these systems some or all of the symbols for units are either short lines carved on rock or metal, indentations made on clay tablets, or strokes of a pen or other similar writing implement, repeated as often as necessary. Thus we find the symbol ||| representing three in the Egyptian engraved hieroglyphic and later written hieratic texts, and also with the familiar Roman numerals which we still see on clocks. We find the same thing with the Cretan and the early Greek Attic numerals, with the stick numerals of ancient China, in texts from the early Indus Valley culture, and with the Kharoṣṭhî script of North–West India. Sometimes we find the strokes written horizontally as with early forms of the Indian Brâhmî and Nâgarî scripts.

In the Babylonian and associated numeral systems we again find the identical principle, this time thin vertical indentations pressed into clay tablets using a stylus with a wedge-shaped cross-section or (even earlier) repeated ◗s made with a stylus of circular section. In these cases three would be represented by ◗◗◗ and then later by ▼▼▼.

Occasionally we find that the unit symbol was a dot rather than a stroke, thus three appears as ••• or as ⁚• in Mayan and Aztec numerals. In all cases the underlying idea is the same as that of tallying. One mark, be it a stroke or a dot or some other symbol, represents a single object or a single unit of measurement. Two identical marks, usually side by side, represent two objects or units. Three identical marks represent three objects or units, and so on. The principle thus has both an additive and a repetitive character: each mark added represents one more object or unit and the marks used in any one system are identical.

Repetition of one and the same symbol is the counterpart of counting in the form 'one–(and)–one–(and)–one–(and)–one . . .', which can hardly be called counting at all. It is certainly not in any sense a system. Repetition of one symbol in a single row rapidly becomes cumbersome, just as much as repetition of the same word. This is true whether it is a matter of notches on a tally stick or strokes or other symbols representing units on paper or clay. Not only does the space needed become too large, but it soon becomes difficult to appreciate just how many units have actually been recorded. We can see this by trying to determine on first sight what number is represented by ||||||||||. Inevitably, we find ourselves having to count the strokes. This does not, however, arise with ||||, which we can immediately read as 'four'.

There is an obvious need to do something to improve the visual impression given by a succession of unit symbols, and the earliest response to this problem, as we suggested in Chapter 2, was to introduce patterns or groupings of symbols. We have already seen examples of very great antiquity, for example, on the palaeolithic wolf bone where the notches are grouped in fives, and on the Philippine sword where the silver nails are grouped in patterns of threes.

It is highly unlikely that the barman serving us with drinks in our local inn would continue to record a long uninterrupted succession of strokes. Instinctively, he is likely to group his marks in fives so that eight pints would be recorded either as ||||| ||| or |||||/|||. He would probably write the five group as ~~||||~~. Here we have just one of the many instances illustrating the extent to which grouping in fives, going back to counting on the thumb and fingers of one hand, has remained ingrained in the subconscious mind of man. Children can also often be seen keeping scores for their various games in this way. We can see the same principle at work with the markings of the bomber aircraft of World War II, where the number of successful bombing raids recorded can be read quickly and easily because the symbols have been painted in rows of ten. Ten is, of course, the base of both our spoken and written numbers and is therefore an obvious and natural grouping, originally stemming from counting on two hands. It does not, however, convey numerical information as immediately as grouping in fives.

We find grouping of unit symbols almost universal amongst the various systems of writing numerals which use the repetitive principle. The Ancient Egyptians, for example, wrote six both in hieroglyphics and in their later hieratic script as |||/|||, though this became 𓏿. Exactly the same arrangement is found in texts from Crete, the Indus Valley, Babylon, and elsewhere. The

Babylonians wrote seven as

or

the Ancient Persians as , and the Aztecs as . It is possible in some cases to connect the method of grouping repeated symbols with ancient systems of counting, that is, with ancient forms of number-words (as we saw in Chapter 2), but it is doubtful if this comparison can be pressed to reliable conclusions. It is also possible to see in the arrangement of repeated symbols in two rows a connection with the method of multiplication by doubling, which we shall discuss in Chapter 4. The temptation to do this is strengthened when we discover that eight is represented by in the Egyptian hieratic numerals. We must, however, be extremely careful not to manufacture connections simply because of chance similarities. The most probable reason for writing symbols in two rows was that they were easier to read.

If we think again about counting soldiers by dropping stones in a box as they march by in single file, we can appreciate that the grouping principle is equivalent to starting a new box each time a given number of men has been recorded. Thus, a new box could be started after every ten men. This greatly facilitates the problem of deciding afterwards how many soldiers have been counted as it is necessary only to count boxes and not the individual stones, except when there is an incompleted box left at the end of the count. In a similar way, after an expensive night at our local inn, the barman will have no difficulty in seeing from IIII IIII IIII II that he has to charge us for seventeen pints, and, if we were faced with the Babylonian number

we would unconsciously interpret it as two threes plus a two in reading the number eight.

There are obvious limitations to the advantages gained by grouping unit symbols in these various ways. To write eight as or IIII III still requires eight strokes of the pen. It demands the same effort and takes up virtually the same amount of space as IIIIIIII.

A second and more satisfactory way of tackling the problem of recording larger numbers is to invent collective symbols so that distinct symbols represent fixed numbers of units. This has the advantage that indefinite repetition of the unit symbol is avoided and grouping is significantly reduced. The only cost, and it is a comparatively small one, is that the meanings of these symbols have to be learned. This process of inventing new symbols is known technically as cipherization, and, as with grouping, it is to be found in all numeral systems, however ancient. There is a suggestion of it even in the case of the palaeolithic wolf bone where we can see notches which are twice the length of others. We cannot be certain, of course, that these longer notches represent a collective unit, but there is a temptation to surmise that even in palaeolithic times the need for at least two symbols for numbers had been appreciated. By contrast, on the Ishango bone, although coming from a much later period, there is no suggestion of any symbol other than the simple notch representing a unit, though the notches are quite clearly in groups.

Following the same principles as with number-words, that is, counting on the thumbs, fingers and toes, we would expect to find widespread examples of special collective symbols for five, ten and twenty units, and this is indeed the case. We find it with many surviving tally sticks just as with the various systems for written numerals. Thus, as well as the simple notches for units, symbols for ten in the form of V or consisting of one notch crossing another and making X are often found. One interesting exception, however, occurs in the case of the Indian Kharosthî numerals where the cross symbol, X, represents four even though succeeding collective units are decimal.

On early Sumerian tablets we find the symbol O representing ten D-units, the O being obtained by pressing the round stylus vertically into the clay. In a similar way the stylus with a wedge-shaped cross-section could be made to produce the indentation ◀ representing ten as well as ▼ representing one. So the number twenty-three would be written as OODDD and later as ◀◀▼▼▼. The corresponding Egyptian symbol for ten was ∩, giving III∩∩ for twenty-three, and there are numerous other examples:

..III (Cretan)
ΔΔIII (Greek Attic)
XXIII (Roman)
✧✧ :. (Aztec)

The corresponding situation when counting soldiers by placing stones in boxes would be to revert to having only one box but two different sizes, shapes or colours of stones, one kind for units and another for tens. Each time a tenth soldier passed by, nine stones representing units would be removed and a new kind of stone representing ten would be placed in the box. It does not matter how the stones are arranged or how much they are shaken up in the box afterwards, the resulting record will be the same. Thus, although with the various written numeral systems it was conventional to write the ten symbols together and not to mix them with the unit symbols, the order of the numerals does not affect the actual numbers themselves so long as only the additive and repetitive principles are employed along with the additional cipherization. So an Egyptian could, in theory at least, see the number represented by two ∩s and three strokes as twenty-three even if they were written in the forms ∩∩III, I∩I∩I, ∩III∩, etc. since all give the same total. This is equally true of all numeral systems built up in this particular way. Only the presence or absence of symbols matters; their order is immaterial.

In each of the examples considered so far, the second distinct symbol has been used to represent ten units. This is by no means universally the case. As with systems of number words, we can find several examples of the first collective symbol representing five. The Romans used V for five, and this is still commonplace. The Greek Attic system included Γ, and the Mayan system —. Seven was thus written in these three systems as VII, ΓII and $\overset{\cdot\cdot}{-}$. With the Chinese stick numerals we find six, seven, etc. written as T, ΠΓ and so on, even though five was written as IIIII. Thus the upper horizontal stroke for five made its appearance only for recording the numbers from six to nine. It is of interest to note that in several instances there was a collective symbol representing five units even though the corresponding spoken number-words were purely decimal, the Roman numeral V being an obvious example.

The Development of Numeral Systems

We can appreciate that there is no limit to the possibilities of inventing new symbols, though in some cases writing implements used enforced considerable restraint on the cipherization process. In principle at least, new symbols could be invented to whatever extent was necessary to meet the needs of a particular culture for commercial, astronomical, or other purposes. In Ancient Egypt we find hieroglyphic symbols for one, ten, hundred, thousand,

Table of Egyption hieroglyphic numerals

and hundred-thousand, and at one time there was a symbol for million though this was little used. Cretan hieroglyphic numerals, which like those of Egypt were purely decimal, included symbols for powers of ten only as far as thousand. In other systems the decimal progression is broken in various ways. The Roman numerals, in addition to the symbols

I, X, C, M or CIƆ, CCIƆƆ, CCCIƆƆƆ

for the unit and powers of ten up to hundred-thousand, included the intermediate symbols

V, L or ↓, D, IƆƆ IƆƆƆ

for five, fifty, five-hundred, five-thousand, and fifty-thousand. The inscription on the *Columna rostrata*, where we can see the symbol CCCIƆƆƆ repeated many times, clearly suggests to us that at the time symbols for powers of ten greater than hundred-thousand did not exist, even though there was manifestly a cultural need for them. It is tempting to suggest that CCCCIƆƆƆƆ would easily have been understood by inference from the symbols for ten-thousand and hundred-thousand. However, the numeral system was not extended in this way.

In the case of the intermediate symbols such as V, L, D, and so on, there was no call for repetition. Two fives are equal to ten, hence VV was necessarily written as X, LL as C, DD as M, and so on. We do not have any evidence that the intermediate symbols were repeated.

Columna rostrata

The underlying principle behind simple numeral systems such as that of the Egyptians is that there was a symbol for the unit, then a second symbol representing the base number of the system B, and then further symbols representing the powers of B, that is B^2, B^3, B^4, . . . as far as was culturally necessary. These were repeated, as required, up to nine times until the next higher symbol became available. Thus, in the Egyptian numeral system with its base ten, the symbols

represent (from right to left)

$$1(10^5) + 2(10^4) + 2(10^3) + 4(10^2) + 5$$

which we would write as 122 405. This example highlights the additive principle and also reminds us that with such systems there was no need for a zero. The symbol for ten-thousand is interesting. It appears to take the form of a bent finger and has been quoted as evidence of a link with finger counting. To accept such a hypothesis, however, we would need also to have evidence of a system of Egyptian finger gestures which represents ten-thousand in this particular way. This evidence has not been produced. Evidence of finger-counting is provided nevertheless by an Egyptian carving, showing the measurement of an ell, where the numbers four, five and six are clearly rep-

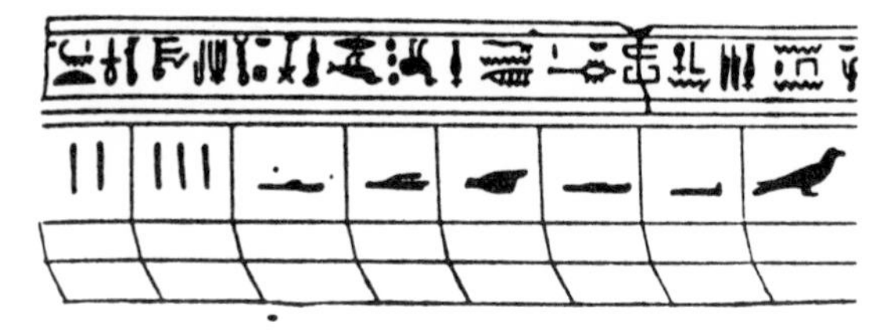

Egyptian carving of measurement of an ell

resented by hieroglyphs derived from counting on one hand. These hieroglyphs are not part of the usual system of numerals.

An example with a base other than ten is provided by Aztec numerals. Here we have dots for units, as with the Mayan numerals, but there is no collective symbol until twenty is reached. Twenty is then represented by a flag. A further symbol in the form of a barbed spike or a feather represents four-hundred, the square of the base twenty, and yet another symbol (described variously as some kind of an epaulette or badge of rank or as a purse) represents eight-thousand, the cube of twenty. So the Aztec numerals

represent one times twenty-cubed, plus eight times twenty-squared, plus two times twenty, plus five, which we would write as 11245.

We should notice that when the base *B* is not ten, the repeated numerals do not transcribe directly into our decimal numerals as they did, for example, with the Egyptian representation of 122 405. The Aztec system was not purely a twenty-based system, however, as we find examples of a spike with barbs on one side only, , representing two-hundred and even examples of spikes with only a quarter of the barbs and with three-quarter of the barbs, and , representing hundred and three-hundred. These aberrations do not invalidate the general conclusion about the nature of the system as a whole.

As when describing systems of number-words, we can use the staircase analogy for written numerals. The Egyptian system is purely decimal, and thus has landings of different orders for the various powers of ten. The Egyptian number

representing one-hundred-and-forty-five requires one large landing-step (of the second order) to reach the , four smaller landing-steps to reach a further , and then five unit steps. By extending the lines, denoting steps and landings, to ground level, the analogy with the written number is complete apart from the way in which repeated symbols are written in two rows. In the cases where units are on the right, the appropriate staircase will go in the reverse direction.

Numeral systems involving the repetition of symbols do not have to be purely additive. Many clocks represent the Roman numeral IIII in the much

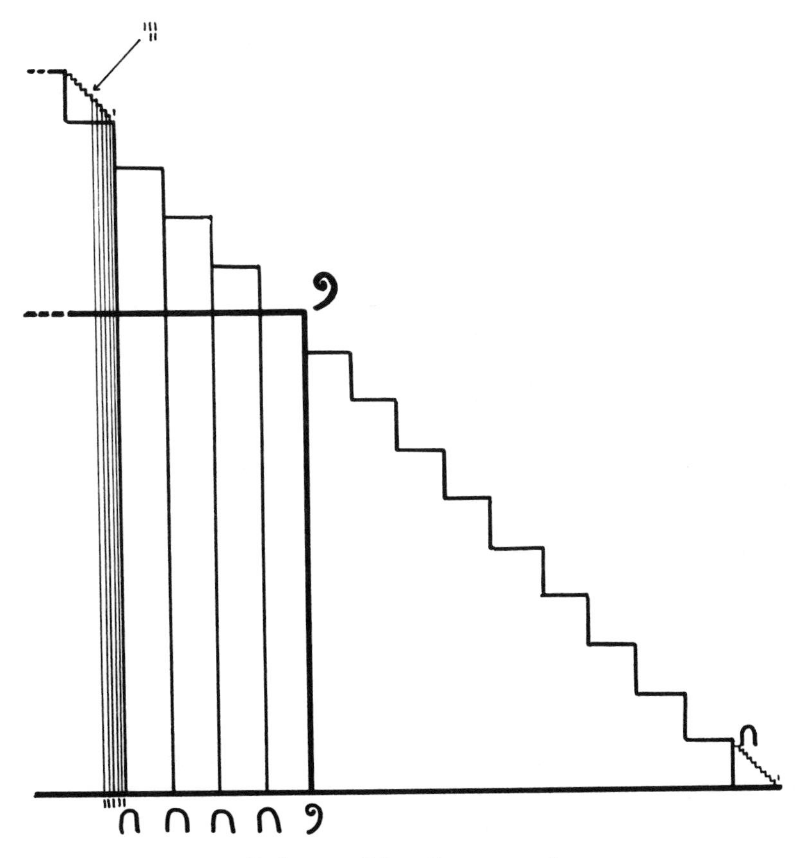

Staircase of Egyptian numerals

later form IV. Here the purely additive principle gives way to the alternative subtractive principle. We find IV meaning 'one less than five', or IX meaning 'one less than ten'. Curiously enough, there is no parallel with the Latin number words *quattuor* and *novem*. Eighteen and nineteen, however, expressed in Roman numerals as XVIII and XVIIII (or later XIX), have the Latin spoken forms *duodeviginti* and *undeviginti* which are clearly subtractive. This underlines the important point that the spoken and written forms of numbers more often than not have an independent history.

The subtractive forms IV and IX achieve a measure of abbreviation at the expense of learning a new convention in which the actual position of the individual symbols becomes important. The freedom found in the Egyptian system to write eleven as ı∩ or ∩ı is now lost. It requires adherence to a strict rule that symbols be grouped in a fixed order, normally in descending order of magnitude, so that any departure from that order means that a different principle, that of subtraction, is involved. Thus, if we see the Roman number MCMLXXIX as the date of copyright of a film or a television programme, we find the descending order of symbols broken twice, and we have to interpret the number as

$$1000 + (1000 - 100) + 50 + 2(10) + (10 - 1)$$

that is, as one-thousand-nine-hundred-and-seventy-nine.

This subtractive principle is not confined to Roman numerals. The Babylonian cuneiform symbol 𒈨 indicates subtraction, and we can find examples of its use so as to reduce repetition. Thus thirty-eight was sometimes written in the form 'forty-minus-two', that is, as

𒐏𒈨𒐖

rather than as

𒌍𒐜

a saving of three symbols.

With the Greek Attic system we find another principle being used, the multiplicative principle. (This parallels the situation in which many number words, such as English *thirty*, are built up.) In addition to the symbols

Ι, Δ, Η, Χ, Μ

for powers of the base as far as the fourth power of ten and the intermediate symbol Γ for five, we find 𐅄 for fifty, 𐅅 for five-hundred, 𐅆 for five-thousand and 𐅇 for fifty-thousand. In each case the symbol Γ has to be interpreted as meaning five times the number represented by the symbol joined to it. So

𐅆ΧΧ𐅅𐅄ΔΓ

means

$$(5 \times 10^3) + (2 \times 10^3) + (5 \times 10^2) + (5 \times 10) + 10 + 5$$

which we write as 7,565. Here, the position of the symbols is once again immaterial, though conventionally the descending order of values was always used. We see this because we can just regard 𐅄, 𐅅, etc. as new and distinct symbols in their own right, in which case only the additive principle need be involved.

If we look at the early Sumerian system where the round stylus was used, we find another example of this multiplicative principle. The six symbols were

D, o, D, ⦸, O, ◎

denoting one, ten, sixty, six-hundred, three-thousand-six-hundred, and thirty-six-thousand. The fourth and sixth symbols are clearly formed by application of the multiplicative principle, ⦸ representing D (sixty) multiplied by o (ten), and ◎ representing O (three-thousand-six-hundred) also multiplied by o (ten). As with the Greek Attic system these symbols were repeated as often as necessary,

O⦸88

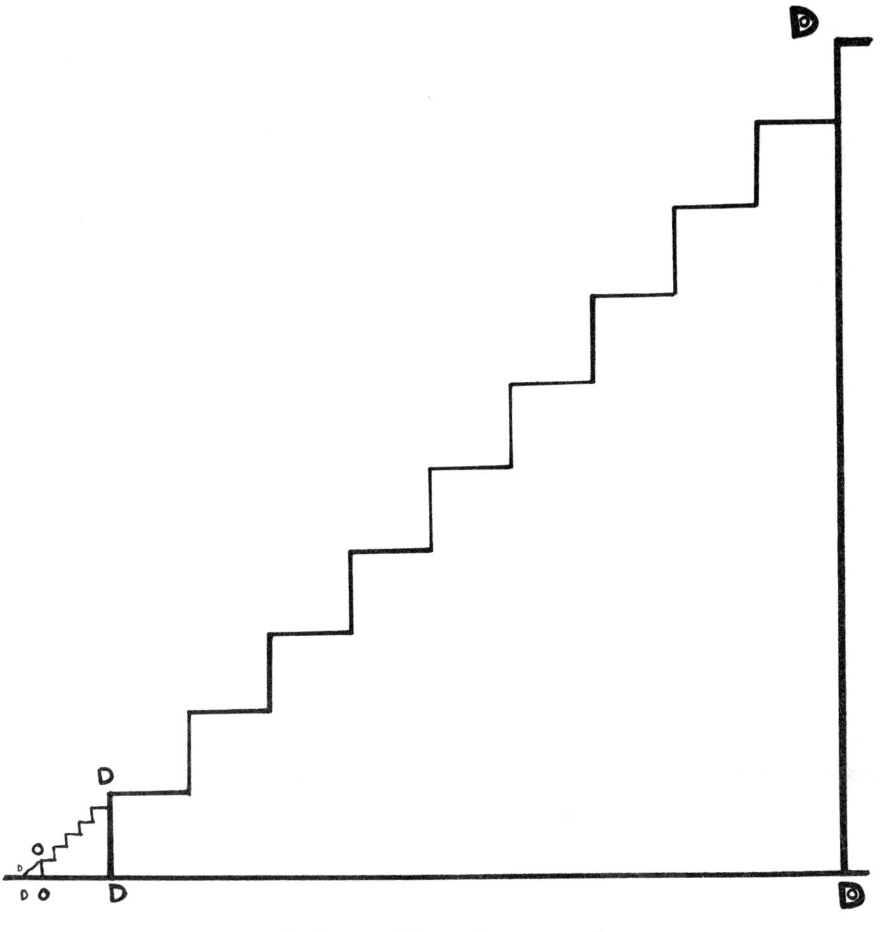

Staircase of Sumerian numerals

for example, representing

$$1(3{,}600) + 1(600) + 2(10) + 2$$

that is, 4,222. The staircase representing this system has landings of value ten, larger landings of value sixty, and so on. Landings of a higher order are reached after five and nine landings of the previous order alternately.

The Sumerian mixture of collective units is of particular interest because it suggests that two cultures having different bases, ten and sixty, for their numerals became merged. This appears to be exactly what happened. The Akkadians had a decimal system and the Sumerians a sexagesimal one, and after the Akkadians became the dominant people the two systems became fused and the mixed system evolved. Later on, it was the sexagesimal base that was to play a crucial role in the development of the Babylonian numerals.

There are advantages to be gained by restricting the total number of symbols used, but such advantages are to some extent offset by other considerations. The more symbols there are, the greater the demand on the memory and the more likely, at least in ancient times, that the result of this would be to restrict understanding of the numerals to the elite. In the Babylonian system, only two different symbols were involved, and writing numbers up to fifty-nine could mean repetitions involving as many as fourteen individual symbols. The compensation was that the demand on the memory for interpreting symbols was minimal. After the number fifty-nine a new and radically different principle was used, as we shall see.

Avoidance of repetition by the invention of more symbols was taken to extreme lengths in the later Egyptian hieratic and demotic scripts, invented for writing on papyrus, wood or pottery. These symbols were in many cases originally abbreviations of hieroglyphic symbols. Their development was made practicable by the much greater freedom allowed by the change in writing materials and the consequent greatly increased speed at which symbols could be written. In the hieratic script we see examples of simple repetition, but in the case of five of the numbers from one to nine the hieroglyphic numerals have been replaced by abbreviated forms. Thus, |||| becomes ⟶ via ⟿; ||||| becomes ⟿ or sometimes just 7; ||||||| becomes ? or ⟶; |||||||| is written repetitively in the form two fours, ⇉; ||||||||| becomes ⟩. (We have already noted that the later hieratic form for six was ⟿.)

This invention of new symbols was extended to multiples of ten, hundred and thousand. After ten-thousand, ?, however, the symbol inventors seem to have given up. From then onwards there was a temporary return to the repetitive principle as far as forty-thousand, and with fifty-thousand and

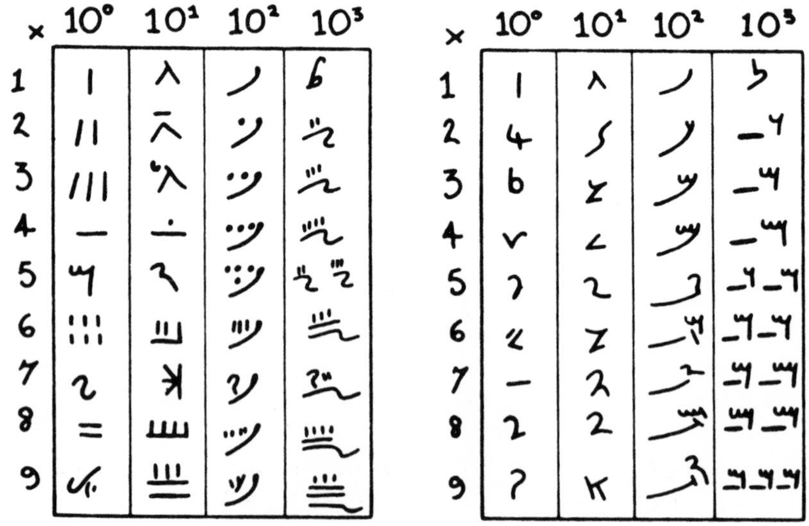

Egyptian hieratic and demotic numerals

upwards the multiplicative principle can be seen, so that forty-thousand appears as but fifty-thousand as . A new symbol appears for the last time with for hundred-thousand, but million, written as , is clearly multiplicative. These hieratic forms were known only to the initiated, that is, very largely only to the priests, and so they became in effect a kind of secret code. It was the demotic forms, which developed from them and could be written even more rapidly, which came into more general use.

Although repetition of symbols was largely avoided by such extensive cipherization, the demand on the memory was considerable; the meaning of an unreasonably large number of symbols had to be learned. Nevertheless, the hieratic forms are especially important in the history of mathematics because important Egyptian mathematical texts, such as the Rhind papyrus in the British Museum, were written in this script. We need to remember this when we read about Egyptian mathematics, since extracts from the mathematical texts are often quoted in their hieroglyphic transcriptions, a fashion set in the early part of this century.

The Evolution of Place-value

It is possible to extend the multiplicative principle much further than with the Greek Attic system. If we combine distinct symbols for one to nine with other symbols for ten and powers of ten, we can effectively write any required number without cumbersome repetition. The principle entails having symbols for units from 1 to $B-1$, where B is the base of the system, and further distinct symbols for B, B^2, B^3, and so on. We find this with certain of the numerals of the Far East. Thus, the Chinese standard numerals have the distinct forms

一, 二, 三, 四, 五, 六, 七, 八, 九

for one to nine, and the forms

十, 百, 千, 萬

for ten, hundred, thousand and ten-thousand. Our 45 657, for example, is written as

四
萬
五
千
六
百
五
十
七

The unit symbols up to and including that for nine are written multiplicatively with those for powers of ten so that, reading from top to bottom we have

$$(4 \times 10^4) + (5 \times 10^3) + (6 \times 10^2) + (5 \times 10) + 7$$

We find a similar situation with the Mayan numerals for recording and computing certain dates. Here the year represents a period of three-hundred-and-sixty days divided into eighteen lesser periods each of twenty days, after which five further days were added to complete the solar year. Larger units consist of twenty and four-hundred years. There was thus a progression by twenties with the one exception of the collective unit representing eighteen smaller units. The ordinary Mayan numerals, • for one and ▬ for five, were used multiplicatively with special hieroglyphs representing the basic unit of one day and the various collective units for twenty and twenty-times-eighteen days, and so on. These hieroglyphs took the form of heads, different forms of heads representing the different collective units. Thus,

is interpreted to mean

$$9(20^3 \times 18) + 12(20^2 \times 18) + 17(20 \times 18) + 4(20) + 17$$

which we write as 1 388 617. This numeral system should be compared with the Mayan number words, discussed in Chapter 2, where eighteen does not play any special role. This is evidence that the number-words were extended to large numbers not for calendar purposes but for religious reasons.

The Mayan numeral system just described can be represented by a staircase with the one irregularity breaking the sequence of landings of different orders occurring with the various powers of twenty. (We do not illustrate this staircase in full here for obvious reasons of space.)

This particular system of Mayan numerals was used almost exclusively for dates. It was, nevertheless, extremely important because, judging by documents which have survived, dates played an important part in Mayan culture, and particularly in religious life. Very large numbers were involved, dates being counted from a purely mythological beginning of their civilization going back thousands of years. The system lacks the special refinement of the Chinese standard numerals as it does not have distinct symbols for units all the way up to its base of twenty but relies on repetition of the symbols for one and five. The Chinese system has the special merit of avoiding repetition altogether.

We have seen how the Babylonians used their two symbols 𒁹 and 𒌋 to write numbers up to fifty-nine. So far, their numeral system was identical in principle to that of the Egyptians. However, they did not represent sixty by

𒌋𒌋𒌋𒌋𒌋𒌋

as we might expect; they adopted an entirely new principle and regarded sixty as a new unit, representing it by 𒁹, already used for one, so that 𒁹 𒌋 𒌋 𒁹 means eighty-one. Here, the vertical wedge on the right represents one in the usual way, and the two ten units make up twenty. The vertical wedge on the left, however, means sixty because it is placed to the left of the tens units. This principle was later extended to higher powers of sixty also. If we look back to our discussion of the early Sumerian numerals, we find that we now have a change of pattern. Instead of landings of higher powers occurring at six-hundred, three-thousand six-hundred, thirty-six-thousand and so on, the new extended system had them occurring at powers of sixty only. Up to and including sixty, of course the staircases representing the two systems are identical.

Until the pure ten-sixty numeral system came fully into being, the multiplicative principle of the old rounded Sumerian numerals was retained, so that for a while six-hundred would be expressed as I<, 'sixty-(times)-ten', just like Ⓓ. In the pure ten-sixty system it came to be expressed as <, though this had to be interpreted very carefully according to the context.

This Babylonian numeral system is the oldest surviving example of the principle of place-value. Clearly, it sets a limit to the need both for the repetition of symbols and for the invention of new symbols. This was crucial for the Babylonians because of the limitations imposed by their writing implements. The base of the Babylonian system was sixty, and we can see the reason for this particular choice of base if we look at their various units of mensuration for commercial and other purposes. These were very largely sexagesimal, because of the many convenient factors of sixty which enabled fractions of a unit, halves, thirds, quarters, fifths, sixths, tenths, twelfths, fifteenths, twentieths, thirtieths and sixtieths, all to be expressed as whole numbers of the next lower denomination. We have the same sort of situation if we omit from the Mayan date numerals the various heads denoting the collective units, or if we write our own geographical and astronomical units omitting the actual symbols for degrees, minutes and seconds and taking the second as the basic unit. Since this last system uses sexagesimal units in historical succession to the Babylonian numerals, the situation would be identical in principle.

We can return to our early discussion of counting soldiers and see the Babylonian ten-sixty numeral system as equivalent to recording numbers using two sizes of pebbles and a collection of containers in a fixed order. The first fifty-nine would be counted by placing five large pebbles and nine smaller pebbles in the first box. As the sixtieth soldier passed by, this box would be emptied and one small pebble put into a second box. These two boxes allow us to record a number up to and including three-thousand-five-hundred-and-ninety-nine. The next soldier would require us to start off a third box, and so on. As long as we keep the boxes in a fixed order, we retain a correct record of the number counted. If the order is changed, however, the information conveyed is radically altered. Of course, we could actually name our boxes 'units', 'sixties', 'three-thousand-six-hundreds', and so on. We would then have a situation like the Chinese and Mayan multiplicative systems which are not true place-value systems but a kind of half-way-house, sometimes referred to as named place-value systems.

Obviously, the Babylonian place-value system could give rise to ambiguities of interpretation. We have already implied this when we mentioned writing six-hundred as 𒌋. Sometimes, the symbol 𒁹 for sixty was larger than the unit symbol, but, just as with 𒌋, standing on its own it had to be interpreted purely by reference to its context. We would have little difficulty today if we saw an orange marked just with the price 8 in deciding that its price was eight pence and not eight pounds. In the same way the Babylonians would know whether a particular article weighed eight or eight-times-sixty shekels. There must however be cases where there is doubt, and this suggests that some principle is lacking from the system. So long as we are in a situation where the collective units are specifically named, as with the Chinese standard numerals and the Mayan date system, ambiguity does not arise. Without the names included, ambiguities are inevitable unless there is a symbol which can denote when any particular place is empty. Thus the place-value principle inevitably throws up a requirement for a symbol to represent zero.

A zero symbol was eventually invented in the Babylonian system, but it was used only to denote internal empty places; it was not used at the right of a number to show that there were no units. Thus 𒁹 𒑱 𒐊 could be read as one times sixty-squared, plus zero times sixty, plus five; but an equally valid interpretation could be one times sixty-cubed, plus zero times sixty-squared, plus five times sixty. To represent this second number unambiguously, the zero symbol 𒑱 would have to be written on the right, giving 𒁹 𒑱 𒐊 𒑱 and this the Babylonians did not do. It was just as if our decimal system of numerals allowed us to write 101 rather than 11 so as to distinguish between one-hundred-and-one and eleven, but did not allow us to write 110. We would thus have no way, apart from reliance on the context, in which we could distinguish between eleven, one-hundred-and-ten, one-thousand-one-hundred, and so on.

In later times, when the Babylonian sexagesimal system had been taken over by the Greek astronomers, a small circle was adopted to indicate zero and used along with the Greek alphabetical numerals which did not otherwise need such a symbol. The Mayan system also had a zero which took the form of a shell, the ornamentation of which was often varied. The shells were used both in the middle and at the end of numbers so that, for example,

unambiguously meant

$$12(20^2 \times 18) + 0(20 \times 18) + 3(20) + 0$$

that is, 86,460.

We find a further example of a place-value system with zero, but one which still involves repetition of strokes, in the Chinese stick-numerals. We have already seen how numbers from one to nine were represented in this system. Ten was represented by a single horizontal stroke in the tens place, and multiples of ten up to fifty by additional horizontal strokes so that ≡ meant forty. After forty, multiples of ten were written:

𝍭	𝍮	𝍯	𝍰	𝍱
fifty	sixty	seventy	eighty	ninety

The symbols were then repeated for hundreds and thousands and the general appearance of a large number was one of horizontal and vertical strokes alternating in groups. The zero symbol, which appeared late, was a circle. As an example, the number thirty-two-thousand-nine-hundred-and-seventy, formed from the individual symbols:

𝍢, 𝍪, 𝍨, 𝍯, O

appeared as

𝍢𝍪𝍨𝍯O

These numerals are read very easily and place little or no tax on the memory.

Alphabetical Numerals

The disadvantage of highly cipherized systems was the need to memorize the details of a large number of special symbols and their numerical meaning. Once an alphabet had been invented, however, there was to hand a ready-made set of familiar symbols in a more or less fixed order which could easily be adopted to represent numbers.

The alphabet, which is of North Semitic origin, dates from the second millennium before Christ, and it was one of the Semitic peoples who first used letters of the alphabet to denote numbers. We do not know when this first occurred, but there is evidence from early Hebrew writings that the twenty-two letters of the Semitic alphabet, which had no vowels, were allocated in their conventional sequence to represent the numbers one to twenty-two.

א	ב	ג	ד	ה	ו	ז	ח	ט	י	כ	ל	מ	נ	ס	ע	פ	צ	ק	ר	ש	ת
1	2	3	4	5	6	7	8	9	10	11	12	13	14	15	16	17	18	19	20	21	22

Allocation of Hebrew letters as numerals

Beyond twenty-two, two letters were used so that אא meant twenty-three, אב or sometimes בב twenty-four, and so on. This is the principle used on the number-plates of British cars, and has the advantage that many more symbols are available for a given place than would be the case were the numerals 0, 1, . . . , 9 only to be used.

The Greeks took over this alphabetical system of numerals along with the alphabet itself, which they modified considerably to meet the needs of the Greek language. The Greek alphabet consists of twenty-four letters, including vowels, and AA denoted twenty-five, AB (or sometimes BB) denoted twenty-six, and so on. Although this system was in theory extendible as far as could be needed, it was used mainly for pages of documents or chapters and volumes of books. We may therefore suspect that it was ordinal rather than cardinal in principle, that is, we should take Greek A, B, Γ, . . . as representing first, second, third, . . . rather than one, two, three, and so on.

Once the principle of using letters for numbers was accepted, the way was open for a further important development. It was almost certainly the Greeks who took this step. Instead of allocating letters to numbers in strict sequence, nine letters were allocated in their order to the numbers one to nine, the next nine letters to ten and multiples of ten up to ninety, and the final nine letters in order to hundred and multiples of hundred up to nine-hundred. Capital letters were used in the first instance, but these were replaced by the small letters once these came into general use. This Greek numeral system existed alongside the earlier Attic system for some time though eventually it replaced it. There was, however, an obvious problem. The system required twenty-seven letters and the Greek alphabet had only twenty-four. This was overcome by inserting three archaic letters ϛ, ϙ and ϡ in the standard alphabet in the sixth, eighteenth and twenty-seventh positions representing six, ninety and nine-hundred respectively.

1–9	α, β, γ, δ, ε, ϛ, ζ, η, θ	(6 = ϛ = vau)
10–90	ι, κ, λ, μ, ν, ξ, ο, π, ϙ	(90 = ϙ = koppa)
100–900	ρ, σ, τ, υ, φ, χ, ψ, ω, ϡ	(900 = ϡ = sampi)
1,000–9,000	͵α, ͵β, etc.	

Allocation of Greek letters as numerals

With these twenty-seven alphabetical numerals all the numbers up to nine-hundred-and-ninety-nine could be written using not more than three symbols at any one time. Since the meaning of each symbol was fixed irrespective of its position, this was not a place-value system and hence there was no need for a zero, though, as we have seen, a zero was introduced in astronomical writings. The letter λ, for example, unambiguously represented thirty and there was no possible confusion either with γ (three) or with τ (three-hundred). Beyond ϡϙθ (nine-hundred-and-ninety-nine) the number sequence was extended by putting a small stroke at the bottom left of the letters α to θ; thus one-thousand, two-thousand up to nine-thousand were written as ͵α, ͵β up to ͵θ. The capital letter M was introduced to represent ten-thousand and multiples of ten-thousand were written either above the symbol M or to its left. Twenty-thousand, for example, was written either as $\overset{\beta}{\mathrm{M}}$ or as βM.

Although it was usually possible to distinguish numbers from words without difficulty because of the context, a bar was often put above numbers so that there could be no possible doubt. Occasionally the distinction was made by writing a prime at the end of a number. Thus the number ͵αφη (one-thousand-five-hundred-and-eight) was distinguished from the word αφη (band) by sometimes writing it as ͵$\overline{\alpha\phi\eta}$ and sometimes as ͵αφη′.

The great advantage which this Greek alphabetical system had over most other ancient systems was that it achieved full cipherization using known symbols. Repetition of symbols, such as we find with the Egyptian, Babylonian and other systems, was completely avoided, and there was no need to commit to memory a large number of unfamiliar symbols as was the case with the hieratic and demotic scripts. It was by no means without its disadvantages, however. Although the letters of the alphabet in order would be known to the educated person, it was necessary also to know the association of the letters representing one to nine with those representing their corresponding multiples by ten and hundred, otherwise very extensive multiplication tables would have been required. Multiplication of *ξ* (sixty) by *τ* (three-hundred) was carried out by multiplying ς (six) by *γ* (three) and then multiplying the result *ιη* by (one-thousand) to give M,*η*. In this was it was necessary to be familiar with the multiplication table only as far as ten times ten.

Archimedes, in his work *The Sand Reckoner*, set up a system of numbers based upon M (ten-thousand), the myriad. Numbers below a myriad of myriads (hundred-million) were described as being of the first class. A myriad of myriads thus began the second class. This process was continued until hundred-thousand to the power hundred-thousand was reached. This incredibly large number became the first number of the first period, and the whole process was repeated. In this way, Archimedes estimated that the number of grains of sand 'contained in a sphere as large as that bounded by the fixed stars' (that is, in the universe) was of the order of ten raised to the sixty-third power. His proposal for writing such large numbers involved a kind of named place-value system in which the usual Greek alphabetical numerals would be combined multiplicatively with powers of the octad (myriad-myriad or 10^8). This was, of course, a purely theoretical system never put into everyday use.

The Greek invention of allocating letters of the alphabet to represent units, tens and hundreds was soon taken over by the very people from whom the Greeks had learned the alphabetical principle itself, though just when this happened is very difficult to determine. The Hebrew alphabet of twenty-two letters, however, allowed the representation of numbers on the Greek system only as far as four-hundred-and-ninety-nine. Until comparatively late times the hundreds above four-hundred were written by juxtaposition of ת, the letter *taw* representing four-hundred, with ק, ר, ש, *goph*, *resh* and *sin*, the letters for one-hundred, two-hundred and three-hundred. Six-hundred, for example, was written תר, and nine-hundred תתק. Thousands were represented by starting again at the beginning of the alphabet and placing two dots over a letter to denote that the number which it represented was to be multiplied a thousand-fold. Eventually the problem of writing hundreds above four-hundred was solved by adding to the ordinary twenty-two letters of the alphabet five special forms in use only at the termination of words. This

1–9	א ב ג ד ה ו ז ח ט
10–90	י כ ל מ נ ס ע פ צ
100–400	ק ר ש ת
500–900	ך (final כ) ם (final מ) ן (final נ) ף (final פ) ץ (final) צ

Hebrew numerals including end-forms

almost exactly parallels the way in which the Greeks had added three archaic forms to their ordinary alphabet to make up the required twenty-seven numerals.

The Hebrew alphabetical numerals replaced a much earlier system of ideographs. These ideographs did not allow for very large numbers to be written down at all, and in the Bible we find 'without number' often used rather than 'a large number', as in *Genesis* 41, 49:

> And Joseph gathered corn . . . until he left numbering; for it was without number.

Indeed, counting in large numbers, being not possible for ordinary mortals, was reserved for God. So, in Psalm 147, 4 we find:

> He telleth the number of the stars; he calleth them all by their names.

The use of letters as numerals allowed words to be associated with corresponding numbers whenever they also made sense in the alphabetical numeral system employed. Such an association of numbers with words, and especially with the names of people and places, came to have special significance in the interpretation of religious and other texts, an idea which has persisted with some of the stranger religious sects to this present day. It is common practice to identify this principle with gematria, though it is only a part of gematria which, in the first instance, was primarily the art of devising and reading secret codes designed by making permutations of the letters of words. There are many examples of gematria in the Bible in both the Old and New Testaments. In *Genesis* 14, 14 we read:

> And when Abram heard that his brother was taken captive, he armed his trained servants, born in his own house, three-hundred-and-eighteen . . .

Then, shortly, in 15, 2 we find reference to 'the steward of my house, this Eliezer of Damascus'. It is surely no coincidence that the name Eliezer can be transcribed in the alphabetical numeral system into three-hundred-and-eighteen. Again, to take a New Testament example, in *Revelation* 13, 18 we are exhorted to 'count the number of the beast' which is stated to be 'Six-hundred three-score and six'. 'Nero Caesar', transcribed into Hebrew and interpreted by numerical gematria, yields exactly this number. Other interpretations of the number of the beast abound. Even serious mathematicians took a hand in trying to discover to whom the passage in *Revelation* is intended to refer. Michael Stifel offered a proof that it referred to Pope Leo X. He argued that if the Roman numeral letters L, D, C, I, M and V are retained from *Leo Decimus* and X (ten) is then added, a rearrangement gives DCLXVI. Some years later, after John Napier, the inventor of logarithms and an ardent protestant, had also demonstrated that it could be interpreted to refer to the Pope, a book appeared by a certain Jesuit, Father Bongus, in which there was a similar demonstration that it stood for Martin Luther. In much more recent times, the number of the beast has been interpreted as the Kaiser and later as Adolf Hitler. No doubt, many other possibilities exist. However, as it is possible that the original Greek text of *Revelation* reads 'Six hundred and sixteen' and was altered later, many of these efforts may have been somewhat misdirected.

That numerical gematria has at times been taken seriously by the Christian Church cannot be doubted, since '99' appears at the end of prayers in several manuscripts, and the Greek letters $\alpha\mu\eta\nu$ represent respectively the numbers one, forty, eight and fifty whose sum is ninety-nine. Indeed, in the last hundred years a number of books have been written by churchmen on the interpretation of numbers in the Bible and the principles of numerical gematria have been used to a quite fanciful extent. Although there is a specific system actually involved here, this kind of approach is very closely associated with mystical concepts of numbers, which we shall mention later.

The Greek alphabet was used as a model for other alphabets such as the Coptic, Gothic and Cyrillic, all of which consist of letters very largely derived

Glagolitic	*Numerical value*	*Cyrillic*	*Numerical value*	*Glagolitic*	*Numerical value*	*Cyrillic*	*Numerical value*
Ⰰ	1	а	1	Ⱉ	700	ѡ	800
Ⰱ	2	б	—	Ⱋ	800	щ	—
Ⰲ	3	в	2	Ⱌ	900	ц	900
Ⰳ	4	г	3	Ⱍ	1,000	ч	90
Ⰴ	5	д	4	Ⱎ	—	ш	—
Ⰵ	6	є	5	Ⱏ	—	ъ	—
Ⰶ	7	ж	—	Ⱏ	—	ь	—
Ⰷ	8	ѕ, ꙃ	6	ⰟⰊ	—	ꙑ	—
Ⰸ	9	з, ꙁ	7				
Ⰺ Ⰹ	10	і (ї)	10	Ⱑ	—	ѣ	—
Ⰻ	20	и	8	Ⱑ	—	ꙗ	—
Ⰼ	30	(ћ)	—	—	—	ѥ	—
Ⰽ	40	к	20	Ⱖ	—	ю	—
Ⰾ	50	л	30	Ⱔ	—	ѧ, ѧ	900
Ⰿ	60	м	40				
Ⱀ	70	н	50	Ⱗ	—	ѫ	—
Ⱁ	80	о	70				
Ⱂ	90	п	80	Ⱘ	—	ѩ	—
Ⱃ	100	р	100				
Ⱄ	200	с	200	Ⱙ	—	ѭ	—
Ⱅ	300	т, ꙧ	300				
Ⱆ	400	оу, ꙋ	400	—	—	ѯ	60
Ⱇ	500	ф, ф	500	—	—	ѱ	700
Ⱚ	—	ѳ	9	Ⱚ	—	ѵ, ѵ	400
Ⱈ	600	х	600	—	—	—	—

Glagolitic and Cyrillic alphabets and allocation to numerals

from the corresponding Greek forms. In most cases the order of the Greek letters and their allocation to numbers was carefully preserved. There is one very interesting exception, however, the Glagolitic alphabet whose derivation from Greek letters, although disputed by some scholars, seems highly likely.

The Glagolitic alphabet was invented for the use of Christian missionaries to the Slavs so that the Gospels and other important Christian documents could be translated into the language of the Slavic peoples. Eventually this alphabet was almost but not entirely replaced by the Cyrillic, invented shortly afterwards for the same purpose and exactly following the Greek order. Although it is possible to explain the majority of the Glagolitic deviations from the traditional order of letters on the basis of interpolation of additional letters to accommodate Slavic sounds not present in spoken Greek, not all the deviations can be explained in this way; a number of them still remain a mystery. They also gave rise to some curious anomalies when the Cyrillic script replaced the Glagolitic. In a rendering of *Mark* 1, 13, the length of Our Lord's temptation in the Wilderness is changed in a Cyrillic version from forty days to twenty days. The Glagolitic letter ⰽ (k) stands for forty and in a Glagolitic version of St. Mark's Gospel is used instead of the corresponding number-word as an entirely correct rendering of the original Greek. On being transliterated directly into Cyrillic ⰽ becomes К, the Cyrillic letter for twenty, so a translator was either ignorant of the difference in the strict order of the letters of these alphabets, was excessively careless, or was determined to be slavishly pedantic.

The Greek alphabetical numeral system was the most efficient of the numeral systems not incorporating place-value. Its great advantage over other contemporary systems lay in its being fully cipherized thus avoiding the inconvenient repetitions of the Egyptian and Babylonian systems.

A great many claims have been made for the overriding superiority of place-value over all other principles. Yet we have only to look at he Babylonian system to see that, despite its incorporation of place-value, the numerals were relatively awkward to use because of the excessive repetition of symbols that was involved.

Of course, one of the advantages of repetitive systems is the ease with which addition and subtraction can be carried out. We can immediately see that in the Egyptian system the sum of ||∩ and ||||∩∩ is ||||||∩∩∩. Even when the addition of symbols of the same kind yields ten or more of them, we encounter no real problem. We have to know merely that ten |s are replaced by ∩, ten ∩s by 𓍢, and so on. It is also very simple to carry out doubling and halving which, as we shall see in Chapter 4, were very important in ancient calculation. With a fully cipherized system, such as the Greek alphabetical numerals, we need to learn addition and multiplication tables. Thus, we have to know that β added to ϵ gives ζ, that adding a further θ gives ις, that doubling η also gives ις, and so on. All this is no longer obvious. Nevertheless, the advantages of full cipherization, properly systematized as in the Greek case, are very considerable. It is, however, the combination of full cipherization with the place-value principle which gives us the most efficient system possible.

For a very short while, a decimal place-value system incorporating the first nine Greek alphabetical numerals and **.** for zero did exist. Its appearance was only transitory, however, and it rapidly gave way to the Hindu–Arabic system of numerals.

The Hindu–Arabic Numerals

It is now universally accepted that our decimal numerals derive from forms which were invented in India and transmitted via Arab culture to Europe, undergoing a number of changes on the way. Their history begins with texts from the Indus Valley culture, roughly contemporary with the Babylonian and early Egyptian texts. Although these are still very largely undeciphered, we can find in them examples of numerals written using strokes grouped together in various patterns.

We know quite a bit about the Indus Valley culture. We know that it was a comparatively advanced culture because excavations at Mohenjo Daro and Harappa have revealed cities with wide streets, brick dwellings, tiled bathrooms, swimming pools and a covered system of drainage. With the oncoming of the Indo–European tribes, around the later part of the third millennium before Christ, Sanscrit was introduced together with the spoken system of decimal number-words. We do not know just how the first numerals evolved nor their exact form. We do know, however, that several different ways of writing numbers evolved in India before it became possible for existing decimal numerals to be married with the place-value principle of the Babylonians to give birth to the system which eventually became the one which we use today.

After the temporary conquest of North–West India by Alexander the Great in 326 B.C., the Maurya Empire was established which spread all over India and into certain other parts of Asia. There are inscriptions from this period on stone columns of the reign of King Asoka and also in caves located in various places. These show that there was then an established system of numerals already in being.

We can see how the forms of the early Indian numerals developed by looking at the Kharoṣṭhî and Brâhmî scripts. The Kharoṣṭhî numerals were written from right to left and employed the repetitive and additive principles and, for large numbers, the multiplicative principle as well. One to three were written as in many ancient scripts by using single and repeated strokes. There was then the distinct new symbol X for four, to which we referred earlier. The numbers one to nine were written:

I, II, III, X, IX, IIX, IIIX, XX, IXX

The system then became decimal, the symbol 7 for ten being repeated in the form 3 for twenty, 73 for thirty, and so on. From hundred upwards the system

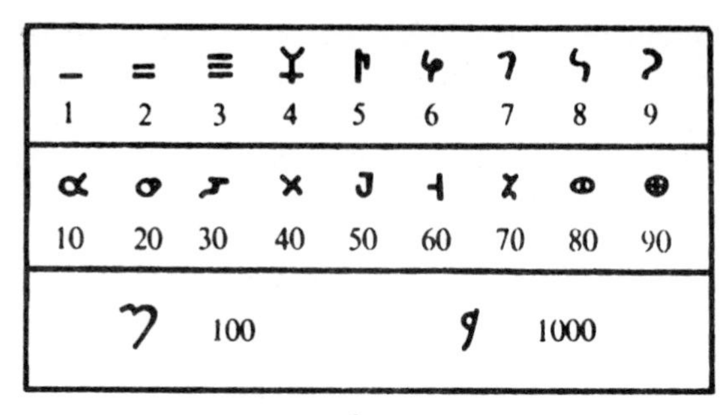

Brâhmî numerals

became multiplicative: hundred was written as , two-hundred as , and so on.

The Brâhmî script is, however, more important than the Kharoṣṭhî because it is from Brâhmî numerals that our own are directly descended. They underwent changes and modifications during the two thousand years or so for which they were in use. The crucial point is that they were used by the Indian astronomers at the time when they first became aware of the place-value system deriving from the Babylonian sexagesimal numerals.

Unlike the Kharoṣṭhî numerals, the Brâhmî forms came to be written from left to right and were exclusively decimal. Two and three were represented by repetitions of the horizontal stroke for one. There were distinct symbols for four to nine and also for ten and multiples of ten up to ninety, and for hundred and thousand. Multiples of hundred and thousand were written using the multiplicative principle: five-hundred was written as , four-thousand as .

The prerequisites for a fully cipherized decimal place-value system are that there should be unique symbols for the numbers one to nine, that the overall system should be decimal, that the place-value principle should be known, and that there should be a zero symbol. The first two of these were provided by the Brâhmî numerals, but the existence of distinct symbols for ten and its multiples, and for hundred and thousand, meant that a substantial part of this system would have to be discarded. Discarding any part of a well-established cultural phenomenon takes place only if there is strong pressure within the culture concerned which can overcome the natural conservatism with which man has always tended to cling to what is familiar. This is especially true if the familiar works well enough in everyday life. In this case the pressure came largely from the need of the Indian astronomers for a numeral system in which very large numbers could be easily represented. This was a period of a great flowering of cultural activity in many areas of learning, but only in astronomy was there the strength of pressure which could change so important a part of the culture as the numeral system. This change must, however, have been influenced favourably by two other aspects of the contemporary scene, the existence of decimal number-words for very high powers of ten and decimal counting boards for making calculations.

We have already mentioned, in Chapter 2, the number-words for very high powers of ten to be found in the *Lalitavistara*. This work, which dates from the first century before Christ, includes decimal number-words up to the fifty-third power of ten. From the same period, we also have a work, the *Anuyogadvara-Sutra*, in which the total number of all living human beings is expressed in terms of decimal denominations occupying twenty-nine *sthana*. Here, the term *sthana*, which we may translate loosely as 'places', does not refer to places in a system of numerals but to their equivalent in written number-words. These number-words of many *sthana* were never put to practical use, but they provided names for the places which were to become an essential part of the decimal numeral system.

If we look at a typical counting board, we see how its columns exactly correspond to the place-value principle. Making a written record of the contents of such a board or at the end of a calculation suggests just such a place-value system as emerged in India in the first half of the sixth century. Counting boards of this type were widely used throughout the ancient world. This meant that the potential for a place-value system for written numbers was by no means confined to India and, as we have seen, such systems did appear elsewhere. The special interest of the Indian system is that it is the earliest form of the one which we use today.

If we are faced with recording an empty column on a counting board then we find ourselves needing a symbol to indicate that the column is empty. It is not good enough to leave a space, though that is what the Indians did in the first instance. The problem is that once the numerals become separated from the board itself, and especially if they are to be copied, the width of the space becomes a crucial factor, and eventually various ambiguities of interpretation can arise.

It is not certain whether the dot or the small circle for indicating that a place was empty was independently invented in India or not. The fact that the Indian astronomers were acquainted with the Babylonian sexagesimal astronomical system, which included the small circle introduced by the Greeks, suggests that this particular symbol, at least, was not an independent Indian invention. The dot was in use, however, well before the small circle, so it may be that the dot was a local invention which was eventually replaced by the small circle taken from Greek astronomical writings. This suggestion is reinforced by examples of the use of the dot from several parts of the Far East.

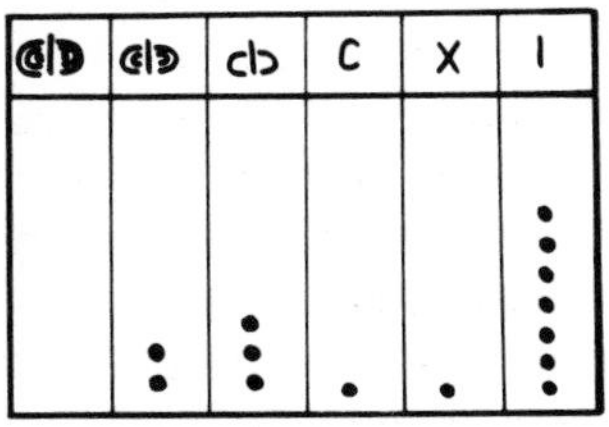

Counting board

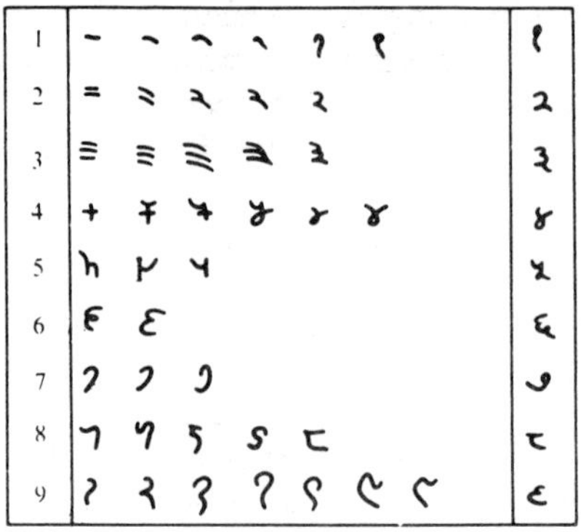

Evolution of Nâgarî numerals

With the passage of time the Brâhmî numerals underwent changes in shape. The separate strokes in the numerals for two and three became joined together and the later forms, known as Nâgarî numerals, show some striking likenesses to our own.

The seventh century saw the founding of Islam by Mohammed and the establishment of the Arab Islamic Empire during the decades immediately following the Prophet's death in A.D. 632. The Byzantine army was defeated, the Persian Empire destroyed and Egypt conquered, and by the early eighth century the Arabs had overrun Spain and even part of Southern France until they were defeated by Charles Martell in 732. To the East they pressed as far as India. Greek, Persian and Indian works were brought to Arab centres of learning, especially Baghdad, and were translated and studied. Towards the end of the eighth century an Indian astronomical textbook making use of the decimal place-value system, the *Siddhânta* of Brahmagupta, was brought to Baghdad and translated into Arabic. This translation had a profound effect upon the history of written numerals both in the Arab world itself and eventually also in the whole of the West. At this time the Arabs were still using the Greek alphabetical numeral system which continued in use, despite the passing in 706 of a law requiring that all official records be kept only in Arabic. The Arabs had no effective numeral system of their own, hence they had little choice but to continue using Greek numerals. Eventually, they did introduce a system on the Greek model using Arabic letters, but this seems to have been fairly short-lived. Indeed, it was as short-lived as the complementary situation, mentioned earlier, where the Greek numeral letters α to θ were used with a dot for zero to form a decimal place-value system.

Although we do have evidence that the Indian or Hindu system was known in the Arab world as early as the middle of the seventh century, the work which was primarily responsible for its dissemination was a booklet written about 825 by the great Arab mathematician al-Khwarizmi. This described both the numerals themselves and their use in calculation. Various spellings of his name have been used, and indeed, although this was the name by which he eventually became known in the West, it is only an abbreviation of a much longer name meaning something like 'Mohammed, father of Dscha'far, son of Musa, from Khwarism', a province of Persia lying to the South of the Aral Sea. Al-Khwarizmi had become familiar with the Hindu system through study of the *Siddhânta*, having been responsible for a revision of this astronomical text.

Knowledge of the Hindu system spread through the Arab world, reaching the Arabs of the West in Spain before the end of the tenth century. The earliest European manuscript in which the Hindu numerals in their modified Arabic forms appear comes from North Spain and dates from 976. Al-Khwarizmi's book was translated into Latin by an Englishman, Robert of Chester, in 1120. He had been visiting Spain to study mathematics and had been fascinated by the system described in the book. This was only the first of several Latin translations by means of which the Hindu–Arabic numerals became known to western scholars. The methods of calculation using these numerals were given the general heading *algorithmus*, from which our word 'algorithm' is derived. Al-Khwarizmi's name has thus been preserved for us, in effect through a confusion between the name of the methods and the author of the principal work through which they were becoming known. *Algorithmus* is just a corruption of the name al-Khwarizmi.

The path of the new numerals in Europe was not to be an easy one despite their obvious superiority for virtually every purpose over the Roman numerals then in general use. We must remember, when looking at history, and especially the history of mathematics, that what seems so obvious to us now was not necessarily obvious to the peoples of times past. For them, a change in numeral system meant not merely learning an entirely new principle for writing numbers but also becoming familiar with strange new symbols which were unlike any encountered before.

Unfortunately the first attempt to introduce the new numeral system was allied to the use of the abacus for calculation. This did not expose their true power. If it was only required to write down the result of a computation carried out with counters on a ruled board, there seemed little point in using symbols which the majority of people would not understand.

This first attempt at introducing the new numerals was due to Gerbert, later Pope Sylvester II. He used special counters, marked with the Hindu–Arabic forms and used in conjunction with a counting board. The principal result of this for the numerals was to introduce confusion about their proper shape. The counters could easily be rotated and anyone unfamiliar with the numerals would then not know whether, for example, the figure ч representing five should be written as ⊂ or as ⊃.

The methods of computation on paper using the Hindu–Arabic numerals came to be widely known through Latin translations of al-Khwarizmi's work. A number of textbooks were written on the 'new arithmetic', and this enabled the power of the fully cipherized decimal place-value system to be more fully appreciated. Even then progress was by no means smooth. There was opposition to the new numerals from Italian bankers, who considered them to be more easily falsified than the Roman numerals. In the commercial world generally their progress was slow, because commercial calculations could be performed with sufficient efficiency on the abacus. The zero symbol itself was a source of difficulty. People found it hard to understand how it was that a symbol which stood for nothing could, when put next to a numeral, suddenly multiply its value ten-fold. This shows how deeply entrenched the additive principle of the Roman numerals was. Compounding this difficulty was the actual word by which the zero symbol came to be known. The original Hindu word *sunya* just meant empty, as did the corresponding Arabic word *sifr*. *Sifr* became *zephirum* in many Latin translations and this was eventually shortened in Italy to *zero*. However, an alternative Latin form was *cifra*, and this, together with the French *chiffre*, came to mean either the zero symbol or any of the nine symbols 1, 2, . . . , 9. These confusions led to overtones of a secret code. Indeed, we still speak of deciphering a secret message. Such overtones heightened popular prejudice against the numerals, and as late as the sixteenth century works on elementary arithmetic using Roman numerals made their appearance. Confusion existed also about the definitive forms of the numerals which became more or less fixed only after the invention of printing.

Throughout this period of uncertainty there had been a number of mathematicians who had strongly supported the new numerals. One of the most prominent and influential of these was the thirteenth-century Italian mathematician, Leonardo of Pisa, also known as Fibonacci. Although born in Italy, he had been brought up in North Africa and had become familiar with the numerals and the Arabic arithmetical methods associated with them. His book, the *Liber Abaci*, is largely devoted to explaining Arabic arithmetic and algebra, and in it he strongly advocates use of the Hindu–Arabic numerals. No other single work contributed more towards the eventual triumph of the new numerals.

It was the mathematicians, rather than the astronomers, who ultimately ensured the almost universal adoption of the Hindu–Arabic numerals. Developments in mathematics, involving an increasing need for elaborate calculation, made adoption of the numerals and the arithmetic associated with them essential. The final triumph over the abacus was so complete that, when the French General Poncelet, himself a notable mathematician, brought a Russian abacus to Paris after having been a prisoner following the Napoleonic campaign against Russia, it was regarded as a very great curiosity. It was forgotten that the abacus had been universally used throughout western Europe not many centuries earlier.

Roman numerals did not disappear as completely as the abacus. They continued to be employed in a number of ways, especially as ordinals, and there must be few western families, even in this twentieth century, who do not have Roman numerals in evidence somewhere around the home. We still see them on the faces of clocks, on the spines of books of more than one volume, and even on our television screens for copyright dates. Nobody, however, would dream of attempting even the most simple calculation using them, except perhaps as a mathematical recreation.

The precise forms which the Hindu–Arabic numerals take is not fixed for all time. We have only to take out our cheque books to see that modern computerized banking methods have brought about some considerable changes. As we move more and more into the computer era, it seems likely that these new somewhat strange looking forms will become the norm, and our present more familiar forms will be reserved for handwritten communications. Perhaps there will come a time in the twenty-first or twenty-second century when historians of mathematics will regard these familiar forms as something of a historical curiosity. Written numbers have their own momentum for change, and their evolution has both a past history and a future potential.

Fractions

The need for some method of representing fractional parts arose in everyday life from early times. If we confine ourselves to counting, we can make do with just the whole numbers, but as soon as we want to divide a single whole into parts or to allocate five sacks of grain equally between three people, we are faced with a situation for which whole numbers are inadequate.

One way of evading fractions is to invent ever smaller units of mensuration, though there are situations in which even this process will not ensure that only whole numbers are needed. Some ancient peoples did avoid fractions in this way, though they were forced to do with only approximations to some quantities they wanted to express. From ancient times, however, certain simple fractions were found to be in such frequent demand that it became necessary to devise special symbols for them. The fractions one-half, one-third, one-quarter, two-thirds and three-quarters arose so often that they came to be known as 'natural fractions', and were given special representations. The Babylonians represented one half as ⊖. This may have been intended to represent a container which had been only half filled just as in mediaeval European manuscripts one-half was often written as +, representing a unit divided in half. The Egyptians represented one-half by ⌓ and one-quarter by ×. This form for one-quarter was probably chosen because it represents the way in which an object might be cút so as to form four equal portions. Early Cretan hieroglyphics include the sign ✓ for one-quarter, which follows the same principle as the Egyptian symbol except that just one of the quarters is actually represented.

The fractions two-thirds and three-quarters have the special property that their denominators are one more than their numerators. In many cultures, we find that not only did they have special symbols, but they also were spoken of as the 'second part' and the 'third part' without any reference to their denominators. This way of referring to them confirms their special character as natural fractions along with one-half, one-third, and one-quarter. There are many other examples which indicate the importance of these natural fractions in ancient cultures. The crucial point, however, is that their widespread use and early date tell us that they were the first fractions to come into everyday usage.

We also find extensive evidence from ancient times of fractions which we write with the numerator one. Such fractions are called unit fractions. The natural fractions one-half, one-third and one-quarter are of this type. In hieroglyphic Egyptian texts they were written by placing the symbol 𓂋 over the number which we call the denominator; thus 𓂋𓎆𓏻 represented one-twelfth. These unit fractions were widely used in Egyptian arithmetic. Indeed, they are to be found in general use in some parts of Africa today.

The Romans adopted a somewhat cumbersome method of representing fractions based on the number twelve even though their counting system was decimal. This method was applied initially to the unit of weight, the *as*, a twelfth of which was known as the *uncia*, a word from which we have derived both English *ounce* and *inch*. We can see the advantage to be claimed for this system of fractions from the following table:

as:	$\frac{1}{12}$	$\frac{1}{6}$	$\frac{1}{4}$	$\frac{1}{3}$	$\frac{5}{12}$	$\frac{1}{2}$	$\frac{7}{12}$	$\frac{2}{3}$	$\frac{3}{4}$	$\frac{5}{6}$	$\frac{11}{12}$	1
uncia:	1	2	3	4	5	6	7	8	9	10	11	12

All the natural fractions of the *as*, and the additional unit fractions one-sixth and one-twelfth as well as certain others, can be expressed as whole numbers of *unciae*. These twelfths of an *as* were then further sub-divided. The *uncia* was divided into twenty-four *scrupuli* and the *scrupulus* into eight *calci*, thus making possible a fractional division as small as a two-thousand-three-hundred-and-fourth part. A special system of notation was devised for these uncial fractions, to which the symbol S was added in the Middle Ages to denote the *semi-as*, eventually contracted to *semis*. Other Roman units, such as the *pes* (foot) were divided in a similar manner.

When we look at the other ancient Mediterranean cultures, the Greek, Egyptian and Babylonian, we find little or no evidence of the duodecimal system for fractions and we are forced to conclude that it was a purely Roman invention. We can trace twelve as a base for certain units of measurement back to the Babylonians, but, as we shall see, they had an entirely different system for writing fractions. Despite their inconvenience, however, Roman fractions continued to be used in Europe, and hence to be taught in textbooks of arithmetic, until at least the thirteenth century.

Fractions, written much as we write them today but without the bar line, appeared in the seventh century in India. The idea of writing one number above another to denote a fraction had been known to the Greeks. It can be found in Greek texts with the denominator sometimes in the upper position and sometimes in the lower one. Often, however, we find the fraction numerals written on the same level with one accent denoting the numerator and two accents the denominator. Again, in the case of unit fractions, it was not uncommon for the denominator alone to be written.

The introduction of the bar line is due to Arab mathematicians, though many Arab manuscripts and even later western Latin manuscripts omit it. Use of the bar presented a problem to the early printers, and it did not become general until the sixteenth century. In many cases, fractions were avoided by

		VALUE	
		As = 1	Uncia = 1
As	I	1	12
Deunx	S=-= S::· [illegible]	11/12	11
Dextans	S== S:: [illegible]	5/6	10
Dodrans	S=- S:· S3 [illegible]	3/4	9
Bes	S= -S- S: [illegible]	2/3	8
Septunx	S- S· [illegible] [illegible]	7/12	7
Semis	S	1/2	6
Quincunx	=-= ::· [illegible]	5/12	5
Triens	== :: ΣΣ [illegible]	1/3	4
Quadrans	=- :· =I [illegible]	1/4	3
Sextans	= : Σ 3	1/6	2
Sescuncia	-L [illegible] [illegible] [illegible]	1/8	1½
Uncia	- · θ [illegible]	1/12	1
Semuncia	Ɛ [illegible]	1/24	1/2
Duella	UU	1/36	1/3
Sicilicus	[illegible] [illegible]	1/48	1/4
Sextula	U	1/72	1/6
Drachma	[illegible]	1/96	1/8
Dimidio sextula	[illegible] [illegible]	1/144	1/12
Tremissis	[illegible]	1/216	1/18
Scrupulus	[illegible]	1/288	1/24
Obulus	[illegible]	1/576	1/48
Bissiliqua	[illegible]	1/864	1/72
Cerates	Z	1/1152	1/96
Siliqua	CIII	1/1728	1/144
Calcus	[illegible]	1/2304	1/192

Roman fractions

means of multiplication. Fractions whose denominators were powers of ten, for example, were just multiplied by a sufficiently large number to remove all the denominators. This was quite a common practice with trigonometrical tables and tables of roots. An approximation to the square root of two, for example, would be written not as $1\frac{414}{1000}$ but as 1 414. This led to incongruities such as the square root of four being written as 2 1000 representing $2\frac{1000}{1000}$. It did, however, allow an easy transition to the adoption of decimal fractions.

In printed texts we come upon fractions written with a solidus as well as those written with a horizontal bar. The practice of using the solidus or angled stroke originated in the eighteenth century in Latin America, and was adopted in England in papers published in 1880. The author was thereupon declared by a contemporary mathematician to have a strong claim to be 'President of the Society for the Prevention of Cruelty to Printers', since other methods of representation had been a constant problem to typesetters. Today there are again examples to be found of the bar being omitted altogether. This is often the case with road signs where a fractional distance of a mile, say one quarter, is just written in the form ${}^{1}_{4}$.

A place-value system does not require any special notation for fractions since the number of places can be extended to the right as far as is required. The Babylonians thus had no need for a special fractional notation though they did make use of a few special symbols for certain natural and unit fractions. Their system of mensuration was sexagesimal, each unit being divided into sixty of the next lower denomination. This meant that their fractions fitted directly into their ordinary numeral system. Sixtieths thus occupied the place immediately to the right of units, and there was no need for any symbols other than those for units and tens already in existence. Their choice of the base sixty meant that for all but a few practical purposes fractional expressions would be finite. In fact, most of their famous tables of reciprocals include only regular numbers, that is, numbers with finite fractional reciprocals.

Decimal fractions were used in China as early as the third century A.D., but the principle does not seem to have spread to other parts of the Far East. They were independently reinvented in the Arab world and in Europe around the fourteenth or fifteenth centuries. Al-Kashi, the astronomer of Samarkand was probably the first mathematician to appreciate that the Hindu numeral system could be easily extended to account for fractions. A Turkish document of the beginning of the fifteenth century has a decimal representation of the value of *pi* taken to sixteen places. In Europe, they made their first appearance in various different forms from about 1350 onwards. Earlier, however, Leonardo of Pisa had provided a prelude to their introduction by writing ordinary fractions on a kind of continuing multiplicative principle. Thus, he would write seven-tenths-and-nine-hundredths as

$$\frac{9\;\;7}{10\;10}$$

the ten under the nine being understood as multiplying that under the seven. This is only a short step from decimal notation, though here the number sequence is reversed.

Credit for the first systematic treatment of decimal fractions as instruments of calculation must go to Simon Stevin of Brughes. His book, *De Thiende* ('The Tenth') appeared in 1582. It was soon translated into French as *La Disme*, and later into English. Stevin's notation was not, however, the notation with which we are familiar today. He identified the units place by putting a zero inside a circle immediately above it. Fractional places were then indicated by putting 1 inside a circle over the tenths place, 2 inside a circle over the hundredths place, and so on. Thus our 354·729 appeared as

$$35\overset{\textcircled{0}}{4}\overset{\textcircled{1}}{7}\overset{\textcircled{2}}{2}\overset{\textcircled{3}}{9}$$

There were many different ways adopted to distinguish between the whole and the fractional parts of a number. Francois Viète, for example, sometimes used a short vertical dash. He was a passionate advocate of the complete decimal system even for astronomical purposes where the Babylonian sexagesimal system was exclusively used and is still in use today. He also expressed the fractional part by writing smaller figures and underlining them. Thus he expressed our 18·275 either as 18|275 or as 18 $\underline{275}$. Other mathematicians used the bracket (or vertical and horizontal bars. Thus we might find 18(275 or 18|$\underline{275}$. A colon was also used.

The use of the comma and dot to indicate the decimal point appeared during the last decade of the sixteenth century, but many different ways of indicating where the fractional places started persisted for another two centuries. On the Continent of Europe it was the comma which eventually triumphed, probably so as to avoid confusion with the continental practice of using the dot to represent multiplication. In England the dot triumphed and confusion with multiplication was avoided by raising the dot above the base line of the numerals. Use of the raised dot did not, however, become a general practice across the Atlantic in America, and, in any case, the dot has tended to descend again for the convenience of typesetters and typists.

The advantages gained by adopting the decimal fraction system are substantial, particularly for calculation. However, even today the metric system of mensuration which goes with it has not yet gained full acceptance. Every British housewife knows this when she finds herself faced in the shops with a confusing mixture of metric and non-metric units and with both ordinary and decimal fractions. Even when the British currency officially became decimal, the authorities retained the old notation ½ for denoting the smallest coin in circulation, half a new penny.

The Derivation of Number-Symbols

We can see that the development of numerals from earliest times went, in the first instance hand-in-hand with that of other forms of writing. Writing had its beginnings in the direct representation of objects by simple pictures. A circle, for example, would represent the sun. This kind of writing could be universally understood irrespective of the language spoken.

Direct representation of objects eventually developed into picture-writing, where the symbols represented ideas connected with objects rather than just the objects themselves. A circle might come to represent light or warmth, either of which would be directly connected with the picture of a sun. At this stage, the universality began to be lost, because the specific ideas represented by symbols could now be a matter of local convention. Thus, the symbol III for three was a direct representation of three cuts on a tally stick or three fingers held upright and could be universally interpreted, and, as we have seen, numerals of this form were common to most ancient civilizations and may well have existed before corresponding number words were spoken.

On the other hand, the Egyptian symbol 𓁨 for million is an example of a local conventional picture symbol and would not have been understood elsewhere. We do not know exactly what this symbol depicted – a god holding up the heavens, perhaps, or a man throwing out his arms in amazement at the thought of the immensity of the number represented. In the same way, the Mayan flag hieroglyph ⊩ for twenty is clearly just a picture, though exactly why this particular symbol denoted twenty is a matter of speculation. An analysis of the association of particular hieroglyphs with numbers in various civilizations makes a fascinating study in which there is ample opportunity for conjecture. In some cases the forms of numerals were related via the words for animals, trees, and other everyday objects to the numbers used in counting them.

Many ancient numerals represent various cuts on tally sticks. We have seen how X was almost certainly just a picture of two crossing notches. This applies widely no matter whether the symbol X be the Roman ten or the Kharoṣṭhî four, though the use of an almost identical symbol in Egypt to mean quarter had, as we have seen, a different derivation. A horizontal cut across notches on a tally stick had the significance of multiplication by ten. We find, for example, instances of ⋇ representing hundred. Later, the top halves of such symbols were used to denote half the value of the whole symbol. As a result, we have the Roman numeral V for five as the upper half of X (ten), and the symbol ↓ for fifty as the upper half of ⋇, a rotated form of ⋇. Although we are more familiar with L for fifty, the alternative ↓ is to be found on milestones surviving to the present day.

Pictorial symbols were sometimes split up in other ways to give new forms of numerals. It is likely that the Roman D for five-hundred is the right-hand-side of CIↃ(thousand) and it is possible that C for hundred was the rest of the symbol. However, there are other theories as to the origins of some of these Roman symbols. One is that they derive from Greek capital letters, such as X and Θ, but this seems a highly unlikely explanation.

In addition to symbols which were clearly representations of objects or ideas derived from objects, there were others derived from geometrical shapes or arrangements of objects in geometrical patterns. Again, such symbols would conform only to localized conventions. The Egyptian ୨ (hundred) may have evolved in this way as its form is very like shapes which were a commonplace decoration of pottery in ancient times. Conventional patterns of objects were used in some civilizations for counting. Some South American Indians, as we saw earlier, lay out objects in geometrical patterns when recording a number. The Chinese stick numerals are pictures of arrangements of actual sticks laid out in patterns to record numbers and make simple calculations. The way in which the dots were arranged in Aztec numerals was probably derived from patterns of stones arranged geometrically during the counting process.

Very early on, the various picture symbols must have become associated with words as their forms became locally conventionalized and simplified. Eventually, each civilization had its own symbols for words, including number-words, and these would be completely meaningless elsewhere except for the repetition of the unit stroke.

Association of symbols with words meant essentially association with sounds of complete words, but eventually this led to symbols for distinct syllables and, still later, to a separate symbol for each individual sound. In this way the alphabetical principle evolved. By now, the symbols bore little or no relation to pictures of objects though, in some instances, we can trace their origin from picture writing as with many of the Hebrew letters.

Some numerals may have been derived from phonetic symbols used at the stage when whole words or syllables were phonetically represented. Certainly, following the invention of purely alphabetical writing by the Phoenicians, there are many examples where a numeral was the initial letter of the corresponding number word. The Greek Attic numerals are one of the most striking examples. 𐅃 (five) is an old form of the first letter of the Greek word *pente*; Δ is the first letter of *deka* (ten), and so on. It seems likely that the Roman C (hundred) is just the first letter of *centum* and M (thousand) the first letter of *mille*. These are the most obvious explanations, at least, but other Roman numerals were certainly not derived on this principle and so it may be that C and M come from CIↃ (thousand), C being the left-hand part of this symbol and M a later form of CIↃ itself. There seems little point in taking the theory of Greek ancestry seriously.

The alphabetical principle of the Greek Attic numerals is entirely different from that of the standard Greek alphabetical numeral system and of systems derived from it, such as the Gothic and Cyrillic. The alphabetical systems which employed letters in conventional order had no connection whatever with the sound or spelling of the corresponding number words. Although the symbols were themselves letters, they now took on an entirely novel significance.

In some cases we find symbols for numbers which are a kind of logical extension of other numerals. This is especially the case with symbols for very large numbers. We find an example with Roman numerals for the higher powers of ten. We must surely wonder, however, why it was that the symbol CCCCIↃↃↃↃ was not invented for million so that the many repetitions of CCCIↃↃↃ (hundred-thousand) seen on the *Columna rostrata* could have been avoided. It seems as logical at least as the eventual use of the outer part of CCCIↃↃↃ to denote multiplication by hundred-thousand and giving ⌈X⌉ as the representation of million.

The source of the Indian Brâhmî forms which were to become our familiar numerals of today is not known with certainty, though the first three numerals bear the obvious relation to tally marks or sticks. It has been suggested that it is possible to derive most of the shapes from earlier Indian picture writing. Unfortunately, much of this early writing remains undeciphered, so it is not possible to have confidence in this hypothesis at present. Other writers have suggested that the original Brâhmî forms derive from Egyptian hieroglyphic or hieratic numerals, but this seems highly unlikely as most of the evidence suggests indigenous origin. Again, it has been suggested that they are derived in some way from early alphabetical letters either on the initial letter principle or on the Greek alphabetical principle. The evidence for an alphabetical origin seems to be the strongest, but this hypothesis awaits further confirmation which will become possible only when the results of more excavations are able to show what forms of numerals were used in the period between the Indus Valley culture and the time of King Asoka. Some of the evidence presented for other hypotheses involves turning the symbols around, chopping off portions and adding various embellishments. If this kind of manipulation is permitted, it becomes possible to provide evidence in support of almost any hypothesis which suggests itself. Many of the arguments which have been put forward cannot be taken seriously.

It is tempting to suggest a definite order of invention such as gestures leading to words, leading to symbols for words, leading to written numerals. Some writers have committed themselves to such a historical schema. No doubt many forms of numerals did evolve in this way, but it seems highly unlikely that all of them did. We must therefore sound a note of caution. Whilst it appears neat and tidy to have a theory of development into which all the evidence can be fitted, it is doubtful if such theories do fit all of it without considerable contriving. Often, important pieces of evidence which simply will not fit are discarded. Clearly, many of the numeral forms had similar or at least parallel histories. Equally clearly, some of them had strange and individual histories which have yet to be discovered.

Chapter Four

Calculating with Numbers

That low man goes on adding one to one,
  His hundred's soon hit;
This high man, aiming at a million,
  Misses a unit.
(Browning)

The different branches of Arithmetic –
Ambition, Distraction, Uglification and Derision.
(Lewis Carroll)

Most of us can carry out the more elementary operations of arithmetic with reasonable efficiency using pencil and paper. Even so, many of us begin to encounter difficulty if we are asked to manipulate numbers in our head, though the problem here may well be one of memory rather than principle. We may well wonder how long this state of affairs is going to last. The pocket calculator is already reducing our practice in calculation. This may save time yet it is important that we do not lose all our skill in arithmetical manipulation. Total reliance of artificial aids would be a major disaster. We cannot guarantee that such aids will invariably be to hand; we must be able to calculate without them. This requires that we should have an understanding of the principles of the calculations which we undertake and that we should maintain the skill to perform them accurately and quickly.

Today there are more-or-less standard methods of elementary calculation taught in school. These have evolved over decades and are often the subject of debate amongst educationalists. They are assumed to suit the vast majority of pupils. But other methods have prevailed in the past. They can suggest alternative approaches even to the most familiar and simple arithmetical operations, and they might well help those at the present time who find special difficulty in understanding and carrying out these operations.

It is our familiar Hindu–Arabic numerals and the methods of calculation which now go with them which have very largely eclipsed earlier ways of coping with the arithmetical demands of everyday life. Vestiges of other methods are, however, still not hard to find. The abacus continues to be used in a number of countries, particularly in the Far East. Calculation on the fingers can still be found amongst several peoples of Africa, the Americas, and Asia. It can also be found in Europe in the Auvergne and amongst various groups of wandering gypsies. Indeed, evidence of the use of fingers for calculation as well as for counting is so widespread that we are led to the conclusion that it must have been a universal or almost universal practice. This is not surprising when we recall that finger gestures for numbers have certainly been practised in every culture at some time of its history. It is only a short step from using the fingers to indicate numbers to using them to perform simple calculations.

From now on we will abandon writing out numbers in words, except where it is important to distinguish between a number itself and the written numeral which stands for it. Numerals should therefore just be read as numbers, that is for example, 8 should be interpreted either as the number eight or as the numeral 8 appropriately according to the particular context.

The most fundamental of the elementary arithmetical operations is addition. This is reflected in that the writers of the great majority of historical treatises on calculation, from ancient times and later, take the ability to perform addition for granted. Few historians of mathematics have devoted space to methods of addition, even when discussing the mathematics of ancient times. This is not to suggest, however, that there has been no variation in the way this operation has been carried out at different times and places.

We can appreciate the fundamental nature of addition by looking at the other three basic operations – subtraction, multiplication and division – and seeing that each of these can be carried out from an additive standpoint. If we are asked to subtract 3 from 9, we shall in all probability give the correct answer, 6, without any conscious reflection. This is because we have become familiar with the result of subtracting any number up to 10 from any larger number. We have memorized what are in effect subtraction tables corresponding to the addition and multiplication tables learned at school. We have, for example, stored in our memory a 'from–9 table' which goes:

1 from 9 is 8
2 from 9 is 7
3 from 9 is 6

and so on. But we can equally well reformulate the question 'what is 3 from 9?' in additive terms and ask 'what must be added to 3 to give 9?'. Indeed, some people tend automatically to do subtraction in this way, and it is a method taught in many schools. It makes subtraction tables redundant.

Multiplication is certainly basically an additive process. If we are asked to multiply 9 by 3 we automatically state the product, 27, from our knowledge of the appropriate multiplication table. We can, however, express the multiplication in alternative additive form by asking 'what is the sum of 9, 9 and 9?'. Of course, this becomes very cumbersome when we have to multiply large numbers together. We would hardly multiply 38 by 17 by adding together seventeen 38s! There is, however, no reason in principle why we should not do so.

Division seems perhaps the most unlikely of the basic operations to be additive. Yet, let us suppose that we have to divide 28 by 7. We can still turn this around and ask 'how many 7s must I add together in order to get 28?'. In fact division has indeed been carried out at certain times by just this principle.

The underlying additive principles which can be discovered in all the basic operations of arithmetic have found expression in variations in the ways in which the operations have been carried out throughout history, although, as we might expect, in the case of addition the variations have been limited.

Addition

With systems of numerals involving repetition, such as the Egyptian hieroglyphic system, new symbols make their appearance only for different powers of the base. Knowledge of addition tables is therefore not necessary. No table is needed in order to add ||| (3) to |||||| (6) to obtain ||||||||| (9), or to add ||∩∩∩𓍢𓍢 (232) to |∩∩∩∩𓆼 (1,041) to obtain |||∩∩∩∩∩∩∩𓍢𓍢𓆼 (1,273). All that we need to do is to juxtapose the corresponding symbols in the two numbers to be added, arranging them if necessary in a conventional pattern. Thus the addition of ||| to |||||| corresponds exactly to bending down 3 fingers and then a further 6 fingers, so ending up with 9 bent fingers in all.

Where addition involves carrying, we have to replace ten of any one symbol by the new symbol for the next higher power of ten, assuming, as in the case of Egyptian numerals, that we are working with 10 as the base of the system. Thus, in reality there are intermediate stages in adding ||||||||∩∩∩∩∩∩𓍢𓍢 (268) and |||∩∩∩∩∩𓍢 (153). We first obtain |||||||||||∩∩∩∩∩∩∩∩∩∩∩𓍢𓍢𓍢 and then, by replacing 10 of the unit strokes by ∩, we obtain |∩∩∩∩∩∩∩∩∩∩∩∩𓍢𓍢𓍢, and finally, by replacing 10 of the hoops by a scroll, we obtain the sum |∩∩𓍢𓍢𓍢𓍢 (421). We do not have any direct evidence to tell us how the Egyptians of ancient times carried out their addition. It is probably fair to assume that the intermediate stages were not written down, at least not by the adept mathematician, though we cannot be certain that they were not required in the classroom. We should remember too that the Egyptian mathematical texts which have come down to us were written in the hieratic script where the simple visual principle of addition did not apply.We do know, however, that intermediate sums were written down in other civilizations – in India, for example.

If we look at addition using Roman numerals we have III added to VI giving VIIII, CCXXXII added to MXXXXI giving MCCLXXIII, and CCLXVIII added to CLIII giving CCCCXXI, neglecting subtractive forms such as IX instead of VIIII. These additions correspond to those which we discussed at the beginning of the preceding paragraph. The situation is, in principle, exactly the same as with addition in Egyptian hieroglyphic numerals except that we have intermediate collective units within an otherwise purely decimal system: V for five units, L for five tens, and so on. Here the addition of III to VI corresponds to first bending down the thumb and all the fingers of one hand, equivalent to the V, and then bending down 3 fingers and a further one finger on the other hand, equivalent to the III and I, making 9 fingers bent down in all. Much more interesting is the fact that addition with Roman numerals exactly corresponds to working with an abacus in which each decimal column is divided into a section containing just one counter used to represent 5 and a second section containing 4 counters representing ones.

It is quite obvious when adding either on the fingers or by juxtaposing like symbols that the same sum is obtained no matter which of two numbers is recorded first. Thus, IIII added to II inevitably leads to the same sum as II added to IIII. Today we express this property by saying that for any two numbers a and b,

$$a + b = b + a$$

We call this the commutative property of addition, a modern term although it expresses something with which man has been familiar from ancient times. Because addition is commutative, it is not necessary to memorize complete addition tables; we do not, for example, need to remember that 7 plus 5 equals 12 as well as that 5 plus 7 equals 12 – it is sufficient to commit one of these sums to memory and to know that the order in which we take any two numbers to be added does not affect the result.

When we have to add using a fully cipherized, i.e. non-repetitive, system of numerals we are faced with the problem of learning addition tables. For example, using the Greek alphabetical system we have to know that β (2) added to ϵ (5) gives ζ (7), that λ (30) added to π (80) gives $\rho\iota$ (110), and so on. We are no longer in a situation where the numerals can just be juxtaposed except in a limited way – adding together different powers of the base ten is still done by juxtaposition, so that, for example, β added to λ is written as $\lambda\beta$. There was a further complication arising from this Greek notation. There was a need to know that κ (20) added to ν (50) gives o (70) and that σ (200) added to ϕ (500) gives ψ (700), as well as that β added to ϵ gives ζ.

The tables could, of course, be reduced by taking the commutative property into account. Even with this reduction, however, there was a demand not found with the Hindu–Arabic system where, for carrying out each of these three additions, we only have to know that 2 plus 5 gives 7.

The Greeks were able to restrict their tables by calculating wherever practicable using the unit roots of the corresponding powers of ten. Thus κ plus ν could first be reduced to the sum of the corresponding unit roots β and ϵ, and their sum ζ then multiplied by 10 to give o. This method of reducing the need for tables meant that only the sums of the unit roots had to be memorized, but it required knowledge of a table of root numbers so that the symbols for multiples of 10 and 100 could be immediately translated into the corresponding unit root.

Carrying out addition in the Greek manner effectively makes use of a property which we call today the distributive property of multiplication over addition. Expressed generally in modern notation this is written as

$$a(b + c) = ab + ac$$

For addition in Greek alphabetical numerals using the root number principle, a is either 10 or 100 – we saw that for 1000 and above the system begins again with α. For the last example which we considered, we start with κ and ν, think of them as ten times β and ϵ, add β and ϵ (corresponding to adding b and c), and then convert ten times the sum ζ to o. Thus a corresponds to 10 in this instance. Like the commutative property of addition, this distributive property of multiplication over addition has been appreciated and used from ancient times though its abstract formulation is comparatively modern.

One of the advantages of our system of numerals is that reduction to root numbers is automatic since no change of symbol is involved. Thus 200 plus 500 is immediately appreciated as involving just the calculation 2 plus 5; place-value takes care of the rest and ensures that we get the correct sum, 700. We need to know addition tables only up to 9 plus 9, and these tables can be reduced again by making use of the commutative property.

In India, where our present-day numeral system originated, there were a number of variations in the detail of how addition was carried out. First, there was the so-called direct method, which is simply what we are taught in schools today. This involves writing down two numbers one below the other with units below units, tens below tens, hundreds below hundreds, and so on, and then adding in the normal way beginning with the units and taking account of any carries. Thus, using the modern form of these numerals, 635 and 387 are added in the following way:

```
 635
 387
1022
```

or, possibly

```
 11
 635
 387
1022
```

In the second case the carries are indicated by placing 1s above the upper numbers. However, in some cases units, tens, hundreds, etc. were first added independently, the addition appearing as

```
 635
 387
  12
 11
 9
1022
```

A second form of addition, known as the reverse method, began with the highest place. The partial sums are now crucial and must be written down. Thus, the same calculation would be written:

```
 635
 387
 9
 11
  12
1022
```

Some mathematical writers suggested that the sum be placed above the numbers to be added, giving:

102
~~9~~~~1~~2
635
387

Here, addition starts from the left with the hundreds, and a partial sum is cancelled and replaced whenever there is a carry. It is more common to find the sum placed at the top in this way in Arabic works using the Hindu numerals; in India the sum was usually written at the bottom. It is possible that the expression 'adding up' owes its origin to the practice of writing sums at the top.

Exceptionally, we can find examples where two numbers to be added are first written vertically. The partial sums are then placed on the right, and the final summation is performed in the normal way with the result at the bottom. We would then have:

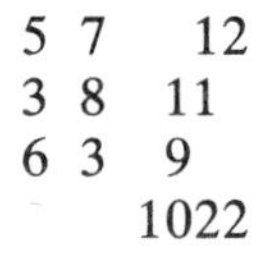

There is no way of being certain here whether addition has started with the units or with the hundreds, though the arrangement of the partial sums suggests that the units were added first. However, this can hardly have been a popular method of writing such calculations. There is clearly more opportunity for mistakes, since care has to be taken to ensure that the partial sums are aligned vertically. Alignment of numbers was sometimes aided by drawing vertical lines. This was quite a common practice amongst the Arabs, but we find it only rarely in the older Indian manuscripts except with one particular method of division.

We shall see in Chapter 5 that addition on an abacus was restricted to two numbers at a time. This was the reason why in ancient times the sum of several numbers was often obtained by stages, two at a time. In adding 237, 512, and 408, for example, the calculations would appear as:

237
512
749
408
1157

It is clear that it was generally appreciated in ancient times not only that addition is commutative but also that it is associative, that is to say, that for any three numbers a, b and c:

$$(a + b) + c = a + (b + c)$$

For the three numbers above this means that we could equally well first add

512 and 408. Doing this and keeping the same order of the numbers gives:

$$\begin{array}{r}237\\512\\408\\920\\1157\end{array}$$

not an arrangement which we would expect to find, though a possible one.

The associative property together with the commutative property means that, given several numbers to add together, it is immaterial in what order we arrange them or how we group them together in additive stages. This makes possible a number of methods for helping with mental addition. Provided that we always add units to units, tens to tens, and so on, we can even rearrange the individual columns independently so that, for example, in calculating the sum

$$\begin{array}{r}23\\16\\77\\94\end{array}$$

we can rearrange an individual column in order to pair together numbers which add to ten. This gives us:

$$\begin{array}{r}\left.\begin{array}{r}97\\73\end{array}\right\}\\\left.\begin{array}{r}26\\14\end{array}\right\}\end{array}$$

Further, such a rearrangement enables us also, in default of other considerations, to add the larger numbers first. Here, the units have been paired on the 'adding to ten' principle, and the tens have been rearranged in descending order of magnitude. We would normally do this mentally.

There is one particular advantage to be gained by the widespread ancient practice of restricting addition to the sum of two numbers. This is the ability to carry out an immediate check by subtraction. It may be that it was this as much as the parallel with abacus computation that encouraged this restriction. The sum of 635 and 387, already obtained as 1022, can be immediately checked by subtracting either of the original numbers from their sum to ensure that we do in fact get the other. Where the calculation is written down, as in the form

$$\begin{array}{r}635\\387\\1022\end{array}$$

it is more convenient to subtract upwards so that 635 is obtained by subtracting 387 from 1022. This facility is lost when several numbers are added simultaneously. Checking by subtraction in this way was widely practised in ancient times, so it is understandable that there was a continuing reason for adding by stages in pairs until some other form of check became generally known.

Subtraction

Much of what we have written about addition applies also to subtraction. There are, however, two very important points to remember with subtraction: it is neither commutative nor associative. Thus, $a - b$ is not the same as $b - a$ and $(a - b) - c$ is not the same as $a - (b - c)$.

The first difficulty which arises occurs when a number being subtracted is larger than what it is to be subtracted from. For example, in subtracting the Babylonian cuneiform number 𒌋𒌋𒐊 (25) from 𒐏𒐖 (42), we have to break down one of the tens into ten units before the five units can be subtracted visually. This gives an intermediate stage where 𒐏𒐖 first becomes 𒌍𒁹𒁹𒁹𒁹𒁹𒁹𒁹𒁹𒁹𒁹𒁹𒁹 before 𒌋𒌋𒐊 can be subtracted to give 𒌋𒐛 (17). As with addition, we have no direct evidence as to how such subtractions were conceived in ancient times. However, we must remember that the absence of surviving examples with the intermediate stage written down is not proof that this was never done.

As with addition, a knowledge of tables is essential for cipherized numeral systems, though it is not necessary to have separate tables for subtraction – addition tables are perfectly adequate. This is especially true if subtraction of one number from another is posed in the additive form 'what must I add to . . . to get . . . ?'. Thus, no matter what form of numerals were being used, a table in the form:

$$6 + 1 = 7$$
$$6 + 2 = 8$$
$$6 + 3 = 9$$

and so on, was all that was necessary for subtracting 6 as well as for adding to 6.

Both direct and reverse methods have been used at different times, that is, sometimes the calculation has begun with the units and sometimes with the highest place. There is, of course, no problem here unless there is a need to borrow as when subtracting 216 from 433. Once borrowing is involved there is then the problem of paying back, and this can be done in more than one way. For example, in using the direct method, i.e. starting with the units, when we wish to subtract 216 from 433, we are immediately faced with the need to subtract 6 not from 3 but from 13. To do this we must borrow 10. The most natural way of doing this is simply to allow the 3 in the tens place to be reduced to 2, so that we have in effect:

$$\begin{array}{c} 2_{1} \\ 4\not{3}3 \\ 216 \\ 217 \end{array}$$

This exactly parallels changing a ten symbol into ten unit symbols when using a repetitive numeral system. In fact, we are not really borrowing at all and there is nothing to pay back.

Once we begin to think strictly in terms of borrowing rather than of just replacing 10 by ten units, it becomes natural to think of having something to pay back. This leads us to the alternative method whereby we pay back by adding one to the next place of the number being subtracted. Thus we would have:

433
216

becoming

1
433
2~~1~~6
2

Here, 6 is again subtracted from 13 to give 7, but we now pay back the borrowed 10 by changing the 1 in the tens place of 216 into 2, and we subtract 2 from 3 to obtain one 10 and the same final result as before.

With the reverse method, which was widely used in India, subtraction begins with the highest place. Thus, in subtracting 216 from 433 the first two differences are obtained immediately. At this stage the calculation appears as:

433
216
22

or as

22
433
216

if the differences are placed at the top, as was often the case. In order to subtract the six we now have to take one from the tens of the partial difference already obtained. The final subtraction then appears as:

1
433
216
2~~2~~7
1

or as

1
2~~2~~7
1
433
216

If subtraction is posed in the additive form, for example, 'what must I add to 216 to get 433?', it would seem to be more natural to pay back by adding one to the lower number. Thus for

$$\begin{array}{r} 433 \\ 216 \end{array}$$

assuming that we start with the units, the first question we ask is 'what must I add to 6 to get 3?'. Having written down 7, we can then ask either 'what must I add to 2 to get 3?' or 'what must I add to 1 to get 2?'. Clearly, either is permissible. Historically, the addition method, often called the Australian method involved the former rather than the latter.

In any of the methods there is immediate opportunity for a check. We simply reverse the process, adding the smaller number to the difference obtained to ensure that we get the larger number.

There is another method which has been employed sometimes, especially in mental arithmetic. This makes use of the complementarity principle. Thus, instead of subtracting 8, we first add its ten-complement, 2, and then subtract 10. We can represent this by the expression

$$a - b = a + (10 - b) - 10$$

In the same way, instead of subtracting the numbers 769 (say), we can first add its thousand-complement, 231, and then subtract 1000. Not much advantage is gained unless the number to be subtracted consists mainly of individual digits greater than 5. Even then there is additional possibility of error since the complement which has to be added can be incorrectly calculated. The advantage claimed is based upon the assumption that addition, as well as being the most fundamental of the operations, is also the easiest to carry out. The method reduces subtraction to the working out of complements together with the subtraction of the number one, that is, we need to be able to subtract any of the numbers 1 to 9 from 9 and to subtract 1 from any other single digit.

Multiplication

The simplest of all multiplicative operations is multiplication by the base of a numeral system or by some power of the base. For our own decimal system multiplication by 10, 100, 1000, and so on, just means writing one or more zeros as appropriate on the right of the number being multiplied or, in the case of a number involving decimal fractions, moving the decimal point the appropriate number of places to the right.

With Egyptian hieroglyphic numerals and other systems on the same principles, multiplication by powers of the base just meant substitution of symbols. Thus, multiplying |||∩∩ (23) by ∩ (10) involved the substitution of ∩ for | and 𓍢 for ∩, giving ∩∩∩𓍢𓍢 (230). Multiplication of the same number |||∩∩ by 𓆼 (1000) would have involved the substitution of 𓆼 for | and 𓂭 for ∩, giving 𓆼𓆼𓆼𓂭𓂭 (23000).

Although the Egyptians did multiply by 10 in exactly this way, they mainly used a different principle, that of doubling and, sometimes, halving – the easiest multiplicative calculations apart from multiplication by powers of the base. For our system, doubling requires knowledge of the twice-times table, but for repetitive numeral systems the situation is even simpler. For us, doubling four to get eight involves three individual symbols, 2, 4 and 8, and knowledge by heart of the product required. Doubling 𓏺𓏺𓏺𓏺 to get 𓏺𓏺𓏺𓏺𓏺𓏺𓏺𓏺, however, merely involves writing down twice the number of strokes, that is, an additional stroke for each one already present. No additional principle is involved in doubling 𓎆𓎆 to get 𓎆𓎆𓎆𓎆 or 𓍢𓍢 to get 𓍢𓍢𓍢𓍢. The only problem occurs when a carry is involved. Thus, the Egyptian calculator had to appreciate that twice 𓏺𓏺𓏺𓏺𓏺𓏺 is not 𓏺𓏺𓏺𓏺𓏺𓏺𓏺𓏺𓏺𓏺𓏺𓏺 but 𓏺𓏺𓎆.

Doubling together with, occasionally, halving and multiplying by 10 formed the basis of Egyptian multiplication. Thus multiplication of 𓏺𓏺𓏺𓏺𓏺𓏺𓏺𓎆 (17) by 𓏺𓏺𓏺𓏺𓏺𓎆 (15) was carried out in one of the following ways:

		𓏺𓏺𓏺𓏺𓏺𓏺𓏺𓎆	𓏺 /
		𓏺𓏺𓏺𓏺𓎆𓎆𓎆	𓏺𓏺 /
		𓏺𓏺𓏺𓏺𓏺𓏺𓏺𓏺𓎆𓎆𓎆𓎆𓎆𓎆	𓏺𓏺𓏺𓏺 /
𓏺𓏺𓏺𓏺𓏺𓎆𓎆𓎆𓍢𓍢	≏	𓏺𓏺𓏺𓏺𓏺𓏺𓎆𓎆𓎆𓍢	𓏺𓏺𓏺𓏺𓏺𓏺𓏺𓏺 /

		𓏺𓏺𓏺𓏺𓏺𓏺𓏺𓎆	𓏺 /
		𓏺𓏺𓏺𓏺𓎆𓎆𓎆	𓏺𓏺
		𓏺𓏺𓏺𓏺𓏺𓏺𓏺𓏺𓎆𓎆𓎆𓎆𓎆𓎆	𓏺𓏺𓏺𓏺 /
𓏺𓏺𓏺𓏺𓏺𓎆𓎆𓎆𓍢𓍢	≏	𓎆𓎆𓎆𓎆𓎆𓎆𓎆𓍢	𓎆 /

		𓏺𓏺𓏺𓏺𓏺𓏺𓏺𓎆	𓏺
		𓎆𓎆𓎆𓎆𓎆𓎆𓎆𓍢	𓎆 /
𓏺𓏺𓏺𓏺𓏺𓎆𓎆𓎆𓍢𓍢	≏	𓏺𓏺𓏺𓏺𓏺𓎆𓎆𓎆𓎆𓎆𓎆𓎆𓎆	𓏺𓏺𓏺𓏺𓏺 /

The first of these was by far the most common since it involved only doubling. 17 is first shown doubled, giving 𓏺𓏺𓏺𓏺𓎆𓎆𓎆 (34). This is then doubled to get 𓏺𓏺𓏺𓏺𓏺𓏺𓏺𓏺𓎆𓎆𓎆𓎆𓎆𓎆 (68). Once again the previous result is doubled to give 𓏺𓏺𓏺𓏺𓏺𓏺𓎆𓎆𓎆𓍢 (136). On the principle that (as we write it)

$$17 \times 15 = 17 \times (1 + 2 + 4 + 8)$$

the four numbers on the left are then added, the symbol ≏ meaning sum, to give the final result 𓏺𓏺𓏺𓏺𓏺𓎆𓎆𓎆𓍢𓍢 (235). The slanting strokes on the extreme right of the calculation simply indicate which numbers are to be added – in this case all of them.

In the second lay-out, the doubling process is carried out twice, and then the final row shows multiplication of the original number 17 by 𓎆 (10). This just involves changing the symbols. Here the final product is obtained on the principle that (as we write it)

$$17 \times 15 = 17 \times (1 + 4 + 10)$$

so that in this instance the row representing 17-times-2 is not marked for addition. In the third lay-out, 17 is first multiplied by 10 and the result of this is then halved. The final product is then obtained as the sum of 17-times-10 and 17-times-5.

All three of these lay-outs may seem somewhat tortuous to us today, but it is clear that this form of multiplication by splitting up the multiplier into parts, so that for the most part only doubling was required, was highly effective. It was certainly a very ancient and well-tried practice. We have already seen that we have evidence of doubling going back to many millennia before Christ. The Egyptian method later developed into what became known as mediation and duplication, where both halving (mediation) and doubling (duplication) are involved together. This method has survived to this day as a principle in certain high-speed computers. The method of mediation and duplication is sometimes called 'Russian multiplication' because it was especially popular with the Russian peasantry both before and after the 1917 Revolution. The calculation of 17-times-15 now appears (in our notation) as:

17	15
~~8~~	~~30~~
~~4~~	~~60~~
~~2~~	~~120~~
1	240
	255

Here, we successively halve on the left-hand side, ignoring any remainders of one, and correspondingly double on the right-hand side. We then cross out those numbers which have even numbers on their left, and we add together the remaining numbers on the right, in this example 15 and 240, giving us the required product 255.

To see just why this method works, we need to appreciate that the numbers which we are adding on the right represent 15 times unity (the zero'th power of two) and 15 times the fourth power of two, that is, 15×1 and 15×16, and that 16 plus 1 equals 17. What we are effectively doing, therefore, is decomposing 17 into the sum of multiples of powers of two so that it can be expressed as 1×16, plus one. This is seen more clearly if we think in terms of writing seventeen as a binary number, that is, as 10001. This means 1×2^4, plus 0×2^3, plus 0×2^2, plus 0×2, plus 1. The numbers on the right which we eventually add together correspond to the places where 1 occurs in the binary representation of 17.

The Babylonians and the Greeks both made use of multiplication tables. Since the former used a place-value notation with base sixty combined with a decimal repetitive system for numbers up to 59, they needed tables which would enable them to write down products up to 59×59. In fact, they used

μῆκος

α	β	γ	δ	ε	ς	ζ	η	θ	ι
β	δ	ς	η	ι	ιβ	ιδ	ις	ιη	κ
γ	ς	θ	ιβ	ιε	ιη	κα	κδ	κζ	λ
δ	η	ιβ	ις	κ	κδ	κη	λβ	λς	μ
ε	ι	ιε	κ	κε	λ	λε	μ	με	ν
ς	ιβ	ιη	κδ	λ	λς	μβ	μη	νδ	ξ
ζ	ιδ	κα	κη	λε	μβ	μθ	νς	ξγ	ο
η	ις	κδ	λβ	μ	μη	νς	ξδ	οβ	π
θ	ιη	κζ	λς	με	νδ	ξγ	οβ	πα	ϙ
ι	κ	λ	μ	ν	ξ	ο	π	ϙ	ρ

βάθος βάθος

μῆκος

Greek multiplication table

tables which went up to 20 × 20, and then extended these by 10 at a time. Several such tables have been deciphered on surviving clay tablets. There is, for example, an 18-times table which goes from 18 × 1, 18 × 2, and so on, up to 18 × 20. It then proceeds with 18 × 30, 18 × 40, 18 × 50, 18 × 60. This suggests that a product such as [cuneiform] (18) times [cuneiform] (47) would be calculated by looking up the products [cuneiform] times [cuneiform] (7) and [cuneiform] times [cuneiform] (40) in the table, and adding the results.

When we turn to the Greek numeral system with its advantage of full cipherization, we encounter again the same problem which we found with addition. There is no immediate visual connection between, say, β (2) times γ (3), β times λ (30), κ (20) times λ, σ (200) times τ (300), and so on. There was a need to learn sufficient multiplication tables to cope with all these and similar products. However, written Greek tables usually go no further than ι (10) times ι, the practice being to reduce larger numbers up to ϡ (900) to their respective root numbers. Thus κ times λ would be calculated by multiplying their respective root numbers β and γ, and then changing the product obtained, ς (6), to χ (600). The partial products were often recorded in full, so that, 23 × 45 might be set out as:

κ γ
μ ε
ω ρ
ρκ ιε
ϡκ ριε ͵αλε

corresponding, in our notation, to:

20	3	
40	5	
800	100	
120	15	
920	115	1035

First of all, κ and μ are multiplied. The corresponding root numbers are β and δ (4). Multiplying these gives η (8), the root number of the required partial product ω. Next κ and ϵ are multiplied in a similar way: β times ϵ gives ι, which becomes ρ. Then, multiplication by γ yields γ times μ as $\rho\kappa$, and γ times ϵ as $\iota\epsilon$. Summing vertically gives ϡκ and $\rho\iota\epsilon$, and finally, summing horizontally gives ͵$\alpha\lambda\epsilon$, the final result.

Multiplication using Roman numerals presents problems as we can see from the following example. Here we multiply 38 by 43 and set out the computation after the Greek pattern:

XXX	VIII	
XXXX	III	
MCC	LXXXX	
CCCXX	XXIIII	
MDXX	CXIIII	MDCXXXIIII

corresponding, in our notation, to:

30	8	
40	3	
1200	90	
320	24	
1520	114	1634

We have, as on previous occasions, used only the additive form of the numerals, IIII instead of the later subtractive IV, and so on.

There are surviving examples of multiplication tables in Roman numerals, but it is not surprising that most calculation involving these numerals was carried out using an abacus or counting board.

The Hindu–Arabic numerals enabled calculations of all sorts to be written out reasonably easily without undue reliance on tables. This is in contrast to Babylonian calculation which, as we have seen, relied extensively on often very lengthy tables. Many different methods of setting out multiplications using Hindu and Hindu–Arabic numerals have been used at different times and places. Some of these clearly reflect procedures originally carried out on different kinds of counting board. We shall discuss some of the more important of these methods, but by no means all of them. Their description goes back to the early Indian astronomical texts, one of the most famous of which was written in the seventh century by Brahmagupta. In this work, the *Brâhma-sphuta-siddhânta*, he lists a number of multiplication methods, others being described by later writers. Some of the methods are known to be much earlier than the dates of their surviving descriptions.

There was a group of methods which was particularly popular and which involved successively moving the multiplier to the right or to the left after it had been written down above the number to be multiplied. One particular form of this was called the 'door-junction' or 'sliding door' method. Originally, it would have been carried out on a sandboard where erasures and replacements of individual digits presented no problem. Suppose we want to multiply 134 by 15. We first write down the two numbers to be multiplied with the multiplier on the top. We therefore have (in modern Hindu–Arabic numerals):

15
134

We now multiply 4 successively by 5 and 1, obtaining first of all 20, the zero of which is written down immediately below the 5 go give

15
1340

We have to carry 2, and this carry must be added to the product of 4 and 1 to give 6. We therefore now replace the 4, whose task is completed, by 6. This gives

15
1360

We next move the numerals 15, representing the multiplier, one place to the left:

15
1360

and we are ready to multiply the next number, 3. Three times 5 gives us 15, which we have to add to 6 giving 21. We replace the 6 by the 1 of 21, and we have:

15
1310

with a carry of 2. Next, we multiply 1 by 3 and add 2, giving 5. We replace the 3 by 5, and again move the multiplier one place to the left. We now have

15
1510

only the 1 being left of the original numerals 134. We multiply this by 5 to give 5, which has to be added to 5 to give 10. So we replace the 5 of 1510 by 0, and remember the carry of 1. What now appears is:

15
1010

Finally, we have 1 times 1, plus the carry of 1, giving 2. So 2 replaces the left-hand 1 in the lower line and we are left with

15
2010

2010 is the product which we want.

All this may seem very detailed and protracted when written out in the way in which we have had to write it. We must remember, however, that when such a calculation is performed on a dust or sand board, the erasures are very easily made and at no time are more than two rows of numerals involved. The upper row just consists of the multiplier, and the only erasures involved occur when moving this one place at a time to the left. It is in the bottom row that the significant changes have to take place. Of course, any numbers to be carried can be temporarily written in a third row, but this is not essential and was probably not done by experienced calculators.

The method which we have just described was known as the direct method. Much more popular, however, was the reverse method where the multiplier was moved successively from left to right. If we carry out the same calculation by this method, the successive stages are:

15	15	15	15
134	1534	1594	2010

Here, we first multiply 1 by 5, and replace the 1 of 134 by 5, and then multiply 1 by 1, writing 1 on the left of the lower figures. We now move the multiplier one place to the right and continue the calculation.

The reverse method was more popular than the direct method because there was less likelihood of confusion over carrying. When a partial product entailed a carry with another carry into a further place, the alteration needed in the reverse method would be a final one since the multiplication involving that further place had already been carried out. With the direct method, a double carry could affect a numeral which had not already been multiplied, and this could give rise to mistakes.

The reverse method survived the diffusion of Hindu–Arabic arithmetic to western Europe, and it appears in a number of European works, such as the fifteenth-century *Crafte of Nombrynge* to be found in the British Museum. In this work, the multiplier is placed below the number to be multiplied, but the principle is the same as that described in Indian works of the eighth century and earlier and in Arabic works from the ninth century onwards.

As written calculations gradually replaced the use of the sandboard, digits which had to be replaced were simply crossed out and the new digits written above or below as appropriate. Surviving examples enable us to see the whole calculation process at a glance. Thus the multiplication of 134 by 15 would have a layout of the form:

```
  0
  9̸1
 28̸9̸
 1̸5̸5̸0
  134
 15
  15
   15
```

or:

```
  0
 20̸1
 1̸5̸5̸0
  1̸3̸4̸
 1̸5̸5̸5
  1̸1
```

In the second layout, as the multiplier is moved to the right, each of its digits is placed in the top space available to it below the number being multiplied and all digits are crossed out once their work has been completed. This, which looks a little confused at first sight, has the advantage of saving space, but because it is not obvious exactly what is happening there is greater opportunity for mistakes.

Another highly popular method of multiplication involved placing the numbers to be multiplied together above and at the side of a lattice inside which the partial products were recorded. For this reason, the method came to be known as the 'lattice' method, or, more popularly, as the 'gelosia' or 'jealousy' method, because jealous ladies are supposed to peep through lattices at the illicit carryings on of their husbands. The origin of this method is uncertain: it may have been either an Indian or an Arab invention. A lattice of squares is first drawn with as many rows as there are digits of the multiplier and as many columns as there are digits of the number to be multiplied, and each square is then divided into two parts by a diagonal. To multiply 134 by 15 we need a lattice of two rows and three columns. The numerals 134 are then written above the lattice and 15 on its right. We now have:

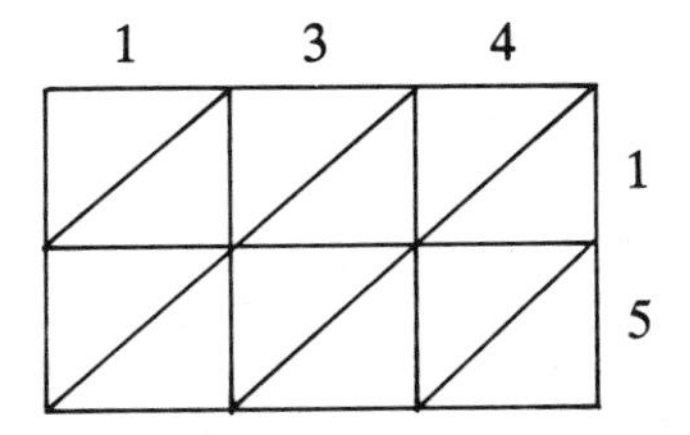

The appropriate partial product for each square is then computed and entered with the tens digit, if any, in the upper triangle and the units digit in the lower triangle. The product, 1×1, the first to be computed, is entered in the top left square corresponding to the column headed 1 and the row with the 1 on the right, and other partial products are computed and entered accordingly. We thus have

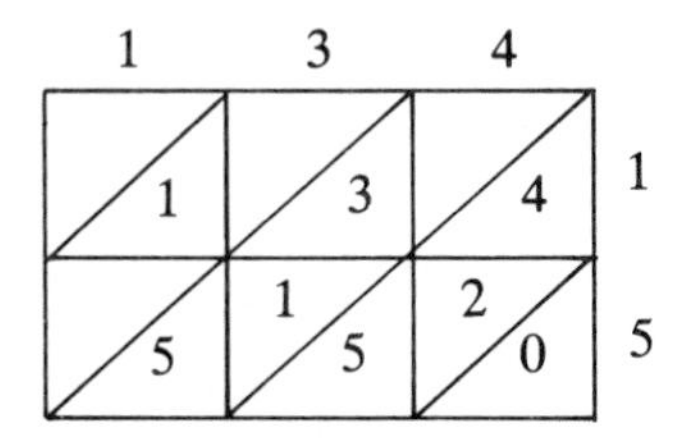

The final product required is then obtained by adding numbers within the lattice diagonally and carrying where necessary, so that we obtain:

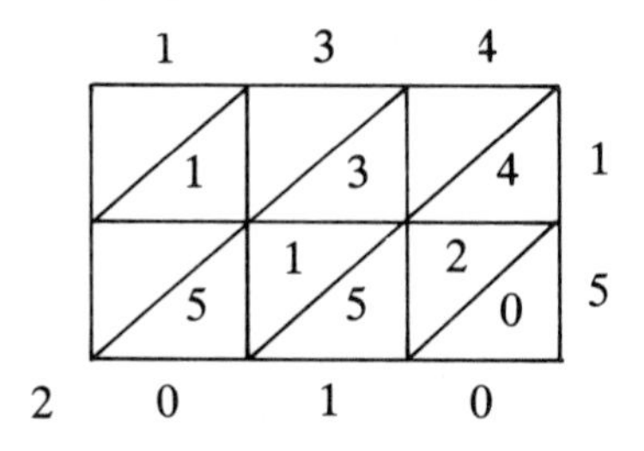

Sometimes, the higher places of the final product would be written up the left side of the lattice. We would then have

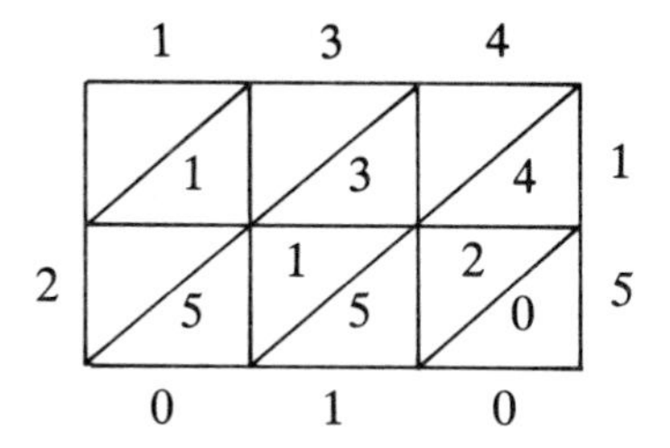

This method of laying out multiplication was widely used in Europe, especially in Italy, and the principle was adapted by the Scotsman John Napier for his calculating rods, invented in 1617, which are the precursors of modern calculating machines.

Yet another method, known variously as the 'cross-multiplication' method, the 'lightning' method or the 'fantastic' method, involves writing down only the two numbers to be multiplied and the final product. Details of all partial products have to be kept in the head or, using finger numbers, on the hands. To illustrate this method we shall consider the product of 3127 by 285.

First of all, we write the numerals in the ordinary way

$$\begin{array}{r} 3127 \\ 285 \end{array}$$

We now begin by multiplying 7 by 5. This gives 35, and we write 5 in the units place, keeping the carry of 3 in our head. We next multiply 2 by 5 and add the carried 3 to give us 13. Remembering this 13, we multiply 7 by 8, giving us 56 to which we add the carried 13, making 69 in all. We write 9 in the tens place, and keep a carry of 6 in our head. At this stage, the written calculation is

```
3127
 285
  95
```

with 6 to be carried into the hundreds place. We now multiply 1 by 5, adding the carried 6 to give 11. To this, we must add 2 × 8, giving us 27, and also 7 × 2, making 41 in all. We write 1 and carry 4 in our head. We now multiply 3 by 5, adding 4 to get 19, to which we have to add both 1 × 8 and 2 × 2, 4, giving us successively 27 and 31. We write 1 in the thousands place, carrying 3 in our head. We have now reached the stage

```
3127
 285
1195
```

with 3 to be carried, having completed the multiplication by 5. We now multiply 3 by 8 and add the carried 3, giving us 27, to which we add 1 × 2, giving us 29 in all. We write the 9 in the ten-thousands place, carrying 2 in our head. This completes the multiplication by 8, leaving us just with 3 × 2, to which we add the carried 2 to obtain 8. We now write 8, and the completed calculation appears as:

```
  3127
   285
891195
```

We can follow the pattern of the whole multiplication process just carried out from the sequence of diagrams below. The various intermediate products are indicated by lines, and these lines are lettered to indicate at each stage the order of carrying out the individual multiplications.

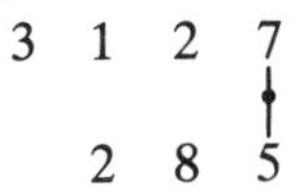

3 1 2 7 / 2 8 5

First stage

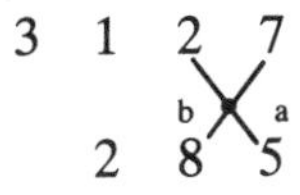

3 1 2 7 / b a / 2 8 5

Second stage

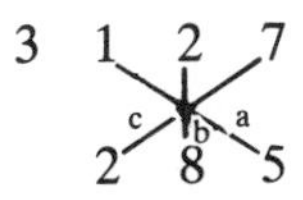

3 1 2 7 / c b a / 2 8 5

Third stage

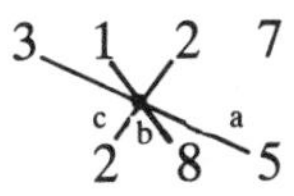

3 1 2 7 / c b a / 2 8 5

Fourth stage

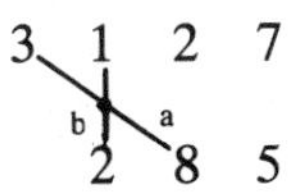

3 1 2 7 / b a / 2 8 5

Fifth stage

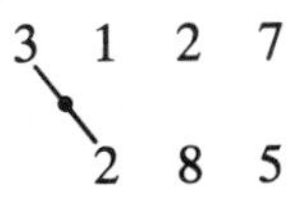

3 1 2 7 / 2 8 5

Sixth stage

If we combine all these diagrams, we obtain the overall pattern below, where each stage is denoted by a blob. This moves from right to left, half a place at a time, as the multiplication progresses.

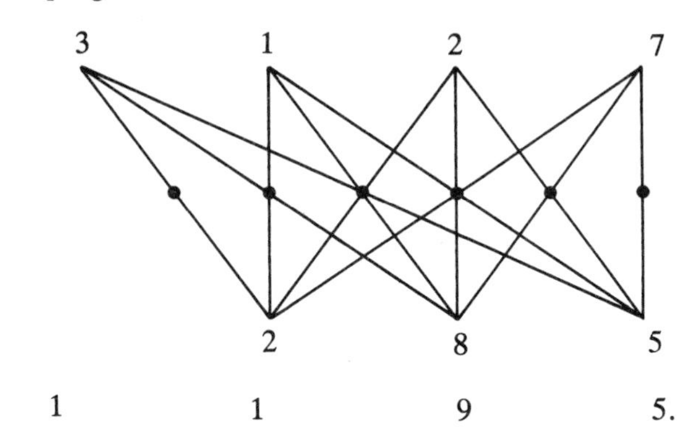

8 9 1 1 9 5.

Although this method might seem very advanced in that skill in multiplication and addition and a good memory are required as well as the ability to add mentally, it is a very old method. In India it goes back to the eighth century at least and it appears from time to time amongst listed methods in a number of early Indian manuscripts as well as in later Arab and European works on arithmetic. Some of this evidence suggests that such multiplications were originally carried out from left to right. When working in this way, the two numbers to be multiplied would have been written with the individual numerals somewhat separated, so that the final product would start under the leftmost numeral and end in the units place. For our example this means that we start by multiplying the 3 by the 2, and write 6 under the blob, that is, half a place to the right of the 3, giving:

```
3 1 2 7
  2 8 5
 6
```

The next stage, 3×8 and 1×2 gives us 26, so we write the 6 below the 2, and change the 6 of the product to 8 to account for the carry of two.

```
3 1 2 7
  2 8 5
 6̸6
 8
```

On a dust board, this would involve erasure of the numeral for 6 and its replacement of that for 8. This is clearly a less convenient way of working, and strict positioning of numerals in the correct place is lost. The final result appears as:

```
3 1 2 7
  2 8 5
 891195
```

Here, for example, the 9 of the product, which is in its ten-thousands place, lies under the hundreds place of the other numbers. This is a highly unsatisfactory situation, and it is not surprising that multiplication from the right was adopted as the rule.

Several of the methods suggested in extant documents involve decomposition of the multiplier either into summative parts, as, for example, with the Egyptian doubling method, or into factors. In some cases subtraction was involved instead of or as well as addition. Thus, to multiply by 126, we first determine its factors. These are 2, 3, 3 and 7. So we multiply successively by these four numbers and add the results. Of course, this involves us in the additional task of calculating factors, and time could be wasted if the multiplier is prime.

If we are to multiply by 29, we would almost certainly do better to multiply by 30 and subtract the number being multiplied from the result. Thus, given the number a to be multiplied by 29, we apply the principle

$$a \times (30 - 1) = (a \times 30) - (a \times 1)$$

that is, the principle that multiplication is distributive over subtraction. We can apply the same principle to multiplication by 98. We just multiply by 100 and also by 2, and subtract. These and similar tricks have been the stock-in-trade of the competent calculator for centuries. Most of them can be traced back to early Indian mathematics.

Many of the methods taught in the early European manuscripts and books were expressly designed to restrict the learning of multiplication tables. It was in any case common practice only to give tables in the abbreviated form

	1	2	3	4	5	6	7	8	9	10
1	1	2	3	4	5	6	7	8	9	10
2		4	6	8	10	12	14	16	18	20
3			9	12	15	18	21	24	27	30
4				16	20	24	28	32	36	40
5					25	30	35	40	45	50
6						36	42	48	54	60
7							49	56	63	70
8								64	72	80
9									81	90
10										100

The arrangement was not necessarily exactly as shown, but there was a clear application of the commutative law

$$a \times b = b \times a$$

to reduce the total number of entries required. This practice extended in Europe to tables of Roman numerals as well as to the Hindu–Arabic numerals, and is also found in the Far East in both China and Japan.

It was, however, found possible to restrict the tables even further so that only products up to 5×5 had to be learned. Methods of multiplication making this restriction possible are described in numerous works from the Middle Ages onwards. For example, in Robert Recorde's *Grounde of Artes*, which appeared in 1542, we find instructions for using the ten-complements for all products in the multiplication table above 5×5. We will discuss a specific case. Consider the product 9×7. We place the two numerals one below the other with their respective ten-complements on the right

$$\begin{matrix} 9 & 1 \\ 7 & 3 \end{matrix}$$

We obtain the tens digit of the product by subtracting diagonally either 1 from 7 or 3 from 9, giving us 6. We obtain the units digit of the product by multiplying the two ten-complements together (always less than 5×5), here giving us 3. Sixty-three is indeed the required product.

This method has been especially popular in the market places of Europe and amongst wandering salesfolk, the gypsies, and similar peoples. To see just why it works, we express it algebraically. Suppose we want to multiply a and b together, both of which are greater than five but less than ten. We write down:

$$\begin{matrix} a & 10 - a \\ b & 10 - b \end{matrix}$$

We now obtain the tens digit of the required product by diagonal subtraction either as

$$a - (10 - b)$$

or as

$$b - (10 - a)$$

both of which give

$$a + b - 10$$

The unit digit is

$$\begin{aligned} &(10 - a) \times (10 - b) \\ &= 100 + ab - [10 \times (a + b)] \end{aligned}$$

The required product is thus obtained as:

$$[10 \times (a + b - 10)] + 100 + ab - [10 \times (a + b)]$$

which simplifies to $a \times b$.

It is not necessary, however, to confine ourselves to this principle only when the numbers to be multiplied lie between 5 and 10. It is, in fact, a special case of a more general method which had its origin in early Indian arithmetic. Whenever one at least of the numbers to be multiplied is near to a power of 10, this general method can be applied. We do, however, have to know the rule of signs, namely that plus-times-plus and minus-times-minus both give us plus, and unlike signs multiplied together give us minus. We give two examples:

$$\begin{array}{rr} 987 & 13 \\ 96 & 4 \end{array} \qquad\qquad \begin{array}{rr} 9968 & 32 \\ 103 & \bar{3} \end{array}$$

In the first of these, we have written the thousand-complement of the upper number and the hundred-complement of the lower number. When we do our diagonal subtraction, we have to take into account that there is a factor of ten between 1000 and 100. Thus we obtain the highest three digits of the required product either by subtracting 40 from 987, or by subtracting 13 from 960. In either case we get 947. We now multiply 13 by 4, giving us 52. The final product can therefore be written immediately as 94752. In the second example, we have written the ten-thousand-complement of the upper number and the hundred complement of the lower number. In the latter case, however, 103 is greater than 100 and its complement is therefore negative. We have indicated this by writing a bar over the 3. This time, the diagonal subtraction has to take into account the factor of one-hundred between 10000 and 100. We first either subtract 32 from 10300, or we add 300 to 9968. Adding 300 is, of course, equivalent to subtracting minus 3 × 100. In either case we obtain 10268. We now multiply 32 by −3, giving us −96. Writing what we have obtained so far gives

$$\begin{array}{rr} 9968 & 32 \\ 103 & \bar{3} \\ \cline{2-2} 10268 & \overline{96} \end{array}$$

We write the bar over 96 to indicate that this is a minus quantity. Subtracting 96 is equivalent to adding 4 and then subtracting 100. We therefore write 04 in place of 96, and subtract 1 from 8, to give us the final product, which we can now write as 1026704.

Our most familiar method today, that of long multiplication, may well have evolved from lattice methods not involving splitting the lattice squares diagonally. Thus, we might have

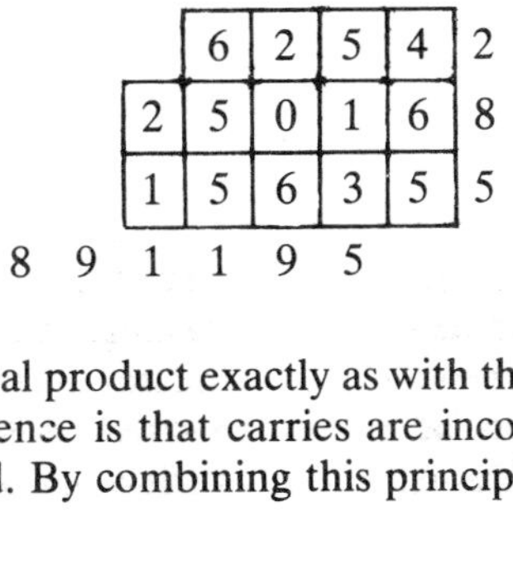

			3	1	2	7	
			6	2	5	4	2
		2	5	0	1	6	8
		1	5	6	3	5	5
8	9	1	1	9	5		

Here, we obtain the final product exactly as with the lattice method described earlier, the only difference is that carries are incorporated when the partial products are computed. By combining this principle with that of moving the multiplier along, we get a rearrangement which makes vertical addition possible instead of diagonal addition:

		3	1	2	7	
6	2	5	4	0	0	2
2	5	0	1	6	0	8
	1	5	6	3	5	5
8	9	1	1	9	5	

This form of layout is sometimes called 'chessboard' multiplication.

It is now a short step to writing down the multiplier beneath the number to be multiplied instead of to the right of the chessboard. However, once this was done it became the fashion to begin, not with the highest place of the multiplier, but with the units place, so that we have

		3	1	2	7
			2	8	5
	1	5	6	3	5
2	5	0	1	6	0
6	2	5	4	0	0
8	9	1	1	9	5

The lattice or chessboard can now be dropped, and we arrive at our familiar form, but with the multiplication of the two units places carried out first:

```
  3127
   285
------
 15635
250160
625400
------
891195
```

The disappearance of the lattice lines, and especially the vertical ones, was very largely the result of the invention of printing, since it is inconvenient for a typesetter to include these. Once the vertical lines disappeared, the only horizontal lines which still made sense were those separating the partial products from the numbers being multiplied above and from the final product below. These survive to the present day.

With the increasing use of decimal fractions, following their introduction by Stevin in the later part of the sixteenth century, and the need for approximate results only in many practical situations, there were advantages to be gained from starting with the highest place of the multiplier. This is because the most significant figures are obtained first, and a calculation can be ended when the places still to be worked out are not required. It is for this reason that the best way of carrying out long multiplication is as in the example below:

```
   3127
    285
-------
6254
25016
  15635
-------
891195
```

We have said nothing about the special case of multiplying a number by itself, that is, calculating the square of a number. Special methods for calculating squares can be found in the *Brâhma-sphuta-siddhânta* and later Indian writings. Two of these methods deserve mention.

The first of these involves a successive process of squaring the last digit, multiplying the rest of the digits by twice this number, and moving the rest of the digits by one place to the left. We will follow through a typical example, finding the square of 124.

First, we square 4, and write 16 above 124, giving

```
 16
124
```

Next, we double the 4, writing 8 underneath the 2 and deleting the 4:

```
 16
12
 8
```

We now multiply 12 by 8, giving 96, and adding in the 1, we write 97 above:

```
976
12
 8
```

We next move the digits 12 one place to the left and delete the 8, which is no longer required:

```
 976
12
```

This completes the first round of operations. Repeating the round with the 2 of 12, we pass through the successive stages

```
1376    1376    5376    5376
12      1       1       1
        4       4
```

Squaring 1 completes the process and gives the answer, 15376. We have followed the instructions working from right to left. It is equally possible, however, to begin on the left.

The other method of squaring is based on a simple principle which we can express algebraically as

$$n^2 = [(n - a) \times (n + a)] + a^2$$

Thus, for two-figure numbers we multiply by the nearest multiple of ten and add the square of the difference between this number and the number which we are squaring. An example will make it clear:

$$38^2 = (36 \times 40) + 2^2$$
$$= 1444$$

In this example, the nearest multiple of ten is greater than the number being squared. If we square 83, however, the nearest multiple of ten is smaller:

$$83^2 = (80 \times 86) + 3^2$$
$$= 6889$$

As the formula is entirely general, it will work for any number, not necessarily only of two figures. However, the great advantage of its application to two-figure numbers is that it enables the squaring to be carried out easily in the head. This is also a very old method, known in Indian from at least the time of Brahmagupta (seventh century). However, the principle was also known in Ancient Greece; it can easily be justified geometrically by a typical Greek proof. We just move the rectangle on the right in the way shown in the diagram. The larger square area n^2 is then obviously equal in area to the rectangle $(n - a) \times (n + a)$ together with the small square a^2.

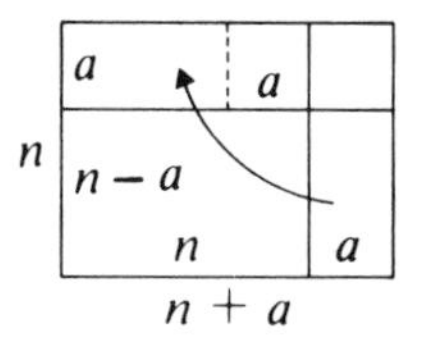

We saw in Chapter 2 how counting on the fingers was a universal precursor of the development of counting systems. Fingers have, however, also been used extensively for calculation and especially for multiplication. This continued even after the advent of knowledge of written methods using the Hindu–Arabic numerals. Leonardo of Pisa (Fibonacci), when commending the use of the new numerals advises those who would master them to become expert also in the use of the fingers. It was not until the sixteenth century that written calculations using the Hindu–Arabic numerals became the norm and the use of the fingers virtually died out in Europe. No doubt the invention of printing and the gradual availability of books in which calculations were printed out in full accelerated the demise of older customs. Apart from the use of fingers at an early stage in the teaching of young children, it is now only in some of the remoter regions of the world and amongst occasional wandering traders that calculation purely on the fingers has survived.

One of the main advantages to be gained from finger multiplication as practised in Europe was that, as with the method in the *Grounde of Artes* described earlier, it reduced the need for knowledge of tables. It required products only up to 5 × 5. In the Middle Ages knowledge of the complete table was considered a substantial feat of memory reserved for the truly expert. It is not surprising that a method which substantially reduced the effort of memory required should survive, if only for a while, alongside calculation on paper using the Hindu–Arabic numerals.

Let us suppose that we want to find the product of 8 multiplied by 6. On one hand we used our fingers to indicate the excess of 8 over 5, that is to say 3, by bending down three fingers. Now on the other hand we indicate the excess of 6 over 5, that is one, by bending down one finger. We now add together the two numbers represented by the bent fingers, 3 and 1, to give 4. This gives us the tens of the product required. Finally, we multiply the two numbers represented by the fingers which we have not bent down, giving 2×4, that is 8. This is the units of the product. This method works for any two numbers from 5 to 9 inclusive. Its justification is very similar to that for multiplication using ten-complements.

Let the two numbers to be multiplied together be a and b, both being in the range 5 to 9. We bend down a-minus-5 and b-minus-5 fingers. We can therefore write the number of fingers still upright as

$$5 - (a - 5) = 10 - a$$

and

$$5 - (b - 5) = 10 - b$$

Adding the numbers represented by the bent fingers gives the tens of the product as

$$(a - 5) + (b - 5) = a + b - 10$$

We can see that we are exactly in the situation of ten-complement multiplication and the rest of the justification follows the algebraic argument which we used in support of that method.

This method of finger multiplication can easily be extended to numbers from 10 to 15. The general rule is now:

bend down a-minus-10 fingers on one hand,
bend down b-minus-10 fingers on the other hand,
add the numbers of bent fingers to give the tens required,
multiply the numbers of bent fingers to give the units required;
we have now obtained the excess of the required product over one-hundred.

The algebraic justification is again very simple. The tens of the product obtained by the method described can be written as

$$(a - 10) + (b - 10) = a + b - 20$$

The units are given by

$$(a - 10) \times (b - 10) = ab - 10a - 10b + 100$$

Overall, then, we have

$$[(a + b - 20) \times 10] + (ab - 10a - 10b + 100)$$

which simplifies to

$$ab - 100$$

This is the excess over 100, so by adding a further 100 we get ab, the product we require.

As an example, we will multiply 12 by 14. We fold down 2 and 4 fingers respectively. Adding gives 6, and multiplying gives 8. The product is therefore 168.

In some cases multiplying to obtain the units figure gives us more than a single digit. For example, if we multiply 7 by 6 or 13 by 14 we get 12 for the units figure in each case. All that we have to do now is to take the units to be 2 and mentally carry the 1, adding it to the tens figure obtained by addition.

Although the method of using the fingers for multiplying just described provides adequately for multiplying by 9, there is an even simpler method involving bending down only one finger and then reading off the product directly from the hands. If we hold up our two hands side by side in front of us with the palms outwards, all that we need to do to read off any product by 9 up to 9×9 is to count from left to right, starting with the little finger of the left hand, and bend down the finger corresponding to the number to be multiplied by nine. The tens digit of the required product is then just the number of fingers upright to the left of the bent finger, and the units digit is just the number upright to its right, fingers and thumbs of both hands being taken into account as appropriate. Thus bending down the index finger of the left hand leaves 3 fingers upright to its left and 6 to its right, giving the product of 4 by 9 immediately as 36.

To see that this method will always work let us suppose that we want to multiply the number a by 9. We bend down the ath finger from the left. This leaves a-minus-1 fingers raised to its left and 10-minus-a fingers raised to its right. Now

$$[10 \times (a - 1)] + (10 - a)$$

just simplifies to $9a$, the required product.

This can be extended to products by 9 from 12×9 to 19×9 by bending down the finger corresponding to the units of the number to be multiplied by 9 and also the finger immediately to its left. The number of fingers raised to the left of the two bent fingers gives the tens digit of the required product, and the number of fingers on their right gives the units, exactly as before. The final product is obtained by adding 100.

Let us try this out on the calculation of 14×9. Holding up our hands as before we bend down the fourth and third fingers from the left, that is, the index and middle fingers of the left hand. We immediately obtain the required product as 126. Algebraic justification of this method is simple. We note that if we are multiplying 10-plus-a by 9, we are effectively obtaining as our answer

$$100 + [10 \times (a - 2)] + (10 - a)$$

which simplifies to

$$9 \times (10 + a)$$

the product which we want.

Methods such as this and other variations of it have been used from time to time, though their origin is obscure. In some cases the details have to be reconstructed from scanty evidence such as an oblique reference in prose or poetry not directly relating to mathematics. Sometimes colloquial phrases associated with specific types of calculation give us a hint of older and forgotten methods. Sometimes we find hints in children's games and surviving folk customs.

We should note particularly that in finger multiplication the developed system of gestures described by Bede and others is not used, though it may have been used to communicate or temporarily record the results of calculations on paper. The multiplication methods, other than the special case for multiplication by 9, were so devised that only numbers up to 5 had to be represented. The obvious and simplest way of doing this was adopted. We have, however, assumed in our examples that numbers would be represented in this simple way by bending the fingers down as with the methods of finger counting. In fact, it is immaterial whether we start with fingers straight and bend them down or start with fingers bent and straighten them. Both methods have been used in finger calculation.

Division

Division has almost always been regarded as the most difficult of the four elementary operations. One of the obvious reasons is that, unless the number to be divided is an exact multiple of the divisor, we are confronted with the problem of either determining a remainder or making an excursion into fractions.

If we look at the way in which the Egyptians carried out divisions not involving fractions, we shall find that they set them out in a way which made them virtually indistinguishable from multiplication. Indeed, instead of saying 'divide 255 by 17' they would say 'reckon with 17 so as to get 255'. Thus, division was posed in multiplicative terms. The hieroglyphic form of this particular division would be as follows (with the modern equivalent shown on the right):

17	1
34	2
68	4
136	8

At this point the scribe would realise that doubling again would lead to a number greater than 255. He would then do some addition on the side to see if the various products obtained so far by doubling could be made to add up to the number being divided. In this case, he would be lucky; the sum of four numbers obtained exactly gives 255. He would then tick the corresponding powers of two to be added, in this case obtaining 1-plus-2-plus-4-plus-8, that is 15, the required quotient. The overall calculation would then appear as:

If we now compare this with the product of 17 and 15 on page 91, we see that it is exactly the same.

This example is, of course, something of a special case in that the number to be divided is an exact multiple of the divisor. There was an especial problem for the Egyptians when this was not so because apart from ($\frac{2}{3}$), their written system of fractions allowed only for unit fractions, that is, fractions whose numerator is unity. Thus they were unable to write the quotient of 19 divided by 8 in a form equivalent to our $2\frac{3}{8}$; still less could they write it in a form equivalent to our 2·375 since their system did not include place-value.

Let us first 'reckon with 8 so as to obtain 19' in the Egyptian manner. We begin by writing

8 1

and then doubling to get

16 2

Here the doubling process stops since twice 16 is greater than the number being divided. We now successively halve the original number, 8, so that below I and II we get the unit fractions $\frac{1}{2}$, $\frac{1}{4}$ and $\frac{1}{8}$:

8	1
16	2
4	$\frac{1}{2}$
2	$\frac{1}{4}$
1	$\frac{1}{8}$

The quotient is now obtained by seeing which numbers in the left-hand column add up to 19, and ticking off the corresponding numbers to their right. This gives (in our notation)

$$2 + \frac{1}{4} + \frac{1}{8}$$

Now let us 'reckon with 12 so as to obtain 31'. We begin, as we would expect:

12	1
24	2

If we now turn to the halving process, we get

6	$\frac{1}{2}$
3	$\frac{1}{4}$

However, trying to halve again, we become involved with fractions on the left since ||| represents an odd number. Also, there is no way in which we can obtain 31 by adding together some or all of 12, 24, 6 and 3. The Egyptian scribe, faced with this situation, had an alternative approach which made use of the special fraction ($\frac{2}{3}$). He would take two-thirds of 12, that is, 8, and then start the halving process. This would enable him to complete this calculation, so that the whole sum would appear as

	12	1
—	24	2
	8	$\frac{2}{3}$
—	4	$\frac{1}{3}$
—	2	$\frac{1}{6}$
—	1	$\frac{1}{12}$

Adding the numbers on the left corresponding to the ticks gives 31, and the required quotient is then obtained (in our notation) as

$$2 + \frac{1}{3} + \frac{1}{6} + \frac{1}{12}$$

Very occasionally it might be possible to obtain an exact quotient using division by 10 as when 'reckoning with 15 so as to get 19'. The calculation then appears in hieroglyphs as

—	15	1
	$1\frac{1}{2}$	$\frac{1}{10}$
—	3	$\frac{1}{5}$
—	1	$\frac{1}{15}$

Here we see division by 10 giving one tenth of 15 as $1\frac{1}{2}$. This is then doubled to get rid of the $\frac{1}{2}$ in the left-hand column, giving one-fifth of 15, namely 3. Finally, one fifteenth of 15 is taken, giving 1, and it then becomes possible to sum on the left to 19 and thus to obtain the quotient required, (in our notation)

$$1 + \frac{1}{5} + \frac{1}{15}$$

Calculations involving unit fractions can often be both tricky and cumbersome. To assist his facility with these fractions the Egyptian scribe had recourse to special tables, notably a table giving the unit fraction representations of the results of dividing two by multiples of three, and also of dividing it by other odd numbers. Direct reference to this table, preserved for us in the Rhind papyrus, enabled the scribe to write down quite complicated expressions such as

$$\frac{1}{42} + \frac{1}{86} + \frac{1}{129} + \frac{1}{301}$$

the result of dividing 2 by 43. A little practical experimentation with a pencil and paper will show that obtaining such a representation of two-forty-thirds is not as easy as it might seem to be.

Like the Egyptians, the Babylonians had recourse to special tables for division. In this case the crucial tables were tables of reciprocals. The division of 25 by 6, for example, was regarded simply as the multiplication of 25 by one-sixth, an expression for the latter in sexagesimal place-value notation being obtained from the appropriate reciprocal table. The reciprocal of (6) was given in the table to be (10), that is, ten sixtieths, and the product of this with (25) was obtained from another table as , that is, 4 units and 10 sixtieths. Of course, only those integers which have as factors 2, 3 or 5, the basic factors of 60, will have finite sexagesimal fractions for their reciprocals. Division by multiples of such numbers as 7 or 11 led in theory to infinite sexagesimal fraction expansions, but in practice the Babylonians always resorted to approximations in such cases.

A number of methods of division have come down to us from early Indian mathematics. Others were rejected by the Arabs and have only recently been rediscovered. One of these, which makes use of the ten-complement principle, deserves special mention.

Suppose that we wish to divide 41 by 9, forgetting about the nine-times table. We turn to the ten-complement of 9, 1, and set up the calculation as follows (in modern notation):

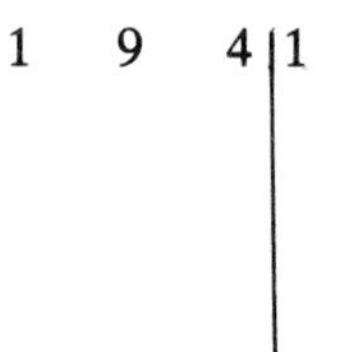

We have written 1 to the left of 9, and drawn a vertical line through 41, separating its first and second places. We begin by writing 4 at the bottom of the calculation. This is the provisional quotient, which may be subject to a subsequent increase. We now multiply 4 by 1, and write 4 beneath the 1 to the right of the line. Adding 1 and 4 gives 5, so we write 5 at the bottom of the calculation, which has now been completed and appears as

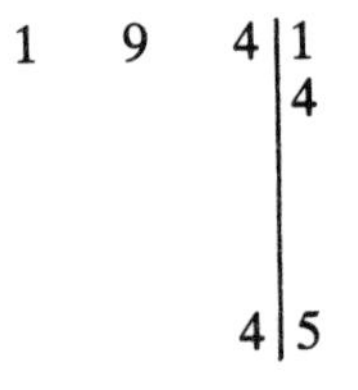

The solution is now read as 'quotient 4, remainder 5'. This is, however, a particularly simple example, so we will now look at one which raises a complication.

We are to divide 3851 by 96. This time, the vertical line is drawn after two places from the right of 3851, and the hundred-complement of 96 is written on its left. We therefore have initially:

4	96	38	51

We begin, as before, by copying the 3 at the bottom of the calculation, and multiplying 3 by 4 to give 12, 12 being written immediately below 85, and adding 8 and 1 to give 9. We now have:

4	96	38	51
		1	2
		39	

The provisional quotient is thus 39. We next multiply the 9 and the 4, using any method limited to tables up to 5 × 5, and write 36 below the 51 on the first free line. Adding, gives us 107. However, this is greater than the original divisor, 96, hence we have to subtract 96, giving the true remainder, 11, and add 1 to the provisional quotient, giving the true quotient, 40. The completed calculation is

4	96	38	51
		1	2
			36
		39	07
		1	96
		40	11

In this case, the provisional remainder was less than twice the original divisor, and obtaining the true remainder and the adjustment to the provisional quotient merely involved subtraction. It is, however, possible for the provisional remainder to be more than twice the original divisor, in which case a separate division has to be carried out. Nevertheless, the operations actually involved are distinctly simpler than long division.

The algebraic justification is simple. We will revert to the first example, which took the form of division of $10a + b$ by c. The quotient found by the method is a, and the remainder is

$$[a \times (10 - c)] + b$$

Multiplying the quotient by the divisor and then adding the remainder should take us back to the number being divided. Indeed,

$$ac + [a \times (10 - c)] + b$$

does simplify to $10a + b$. This justification can be extended to cover larger numbers.

Methods of division from which our familiar modern method is derived go back at least to the fourth century in India. In an eighth-century work, which clearly assumes that the method has been known for a very long time, detailed instructions for long division can be found. First, we are told to ensure that any common factors are removed. We then have to write down the divisor underneath the number being divided as with

```
1750
14
```

Seventeen is then divided by 14 and the quotient, 1, is written either on the right or at the top. On the dust board, the figures for 1 and 7 would then be rubbed out and the remainder 3 substituted for them. Later on, with purely written calculation, 17 would be crossed out and 3 written above it to give (in our numerals)

```
 3
1750      1
14
```

We next have to move the divisor one place to the right:

```
 3
1750      1
14
 14
```

35 is now divided by 14, giving quotient 2 and remainder 7. We therefore write 2 on the right and replace 35 by this remainder, moving 14 into its final position:

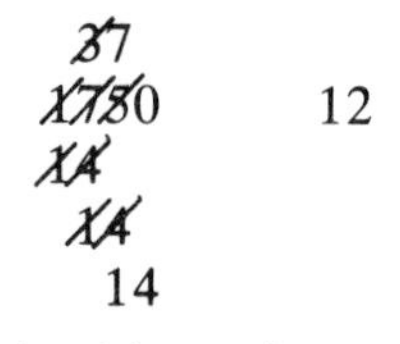

Finally, we have to divide 70 by 14, getting quotient 5 and no remainder, so that we eventually find 125, the overall quotient, on the right.

This method was transmitted to the Arabs, probably about the ninth century, and it subsequently travelled to Europe where it became known as the 'galley' or 'scratch' method of division. In Europe it became usual to fill up the empty spaces when moving the divisor to the right, so that the calculation which we have just discussed would appear as:

```
 37
1750      125
1144
 11
```

The name 'galley method' was probably due to the overall general appearance of such a calculation which could look like a sail. Indeed, in Italy pupils were sometimes encouraged to sketch a picture of a boat in full sail around a completed division. The method was taught in Europe as late as the nineteenth century and is still to be found in schools in some Arab countries.

The modern method of long division, clearly derived from the galley or scratch method, came into being at least as early as the fifteenth century. There were different variations in the positions of both the quotient and the divisor, the former appearing sometimes on the right and sometimes on the top, and the latter appearing on the right before finally settling down on the left as we have it today. Thus, the example above could appear as

```
 125
1750   |14
14
----
 35
 28
----
  70
  70
```

or, later, in either of the forms

```
      125
14 |1750             14|1750|125
    14         or        14
    --                   ----
    35                    35
    28                    28
    ---                  ----
     70                    70
     70                    70
```

The method, known as the 'Austrian' or 'contracted' method depends on making use of the additive principle of subtraction. It has the advantage of requiring fewer lines of writing, but it demands more use of the memory. In dividing 1750 by 14, we would first begin in the usual way:

```
      1
14 |1750
```

Now, however, we write below the 17 that number which, added to 14 times unity, gives 17, that is, 3:

```
      1
14 |1750
      3
```

We now carry on as usual, writing

```
      12
14 |1750
      3
```

and then write in that number which, added to 14 × 2, gives 35, that is, 7:

```
      12
14 |1750
      3
       7
```

Finally, we complete the quotient in the normal manner.

This method has a lot to commend it. In fact, the demand on the memory can be kept to a minimum by working out the partial remainders one place at a time. In our example, there was only one place involved in each remainder. In some documents we find both divisor and quotient written on the right, so that we would have

```
1750|14 = 125
 3
  7
```

The Austrian method can still be found in some European schools, particularly, as we might expect, in Germany and Austria.

Checks on Calculation

The most obvious way to check a completed calculation is to carry out the inverse process. We can check addition by subtraction, subtraction by addition, multiplication by division, and division by multiplication. Many ancient writers recommended that checks should always be carried out in this way. It is, however, often cumbersome, especially with multiplication and division, to use the inverse check and it is not very practicable when more than two numbers are added at a time. It is also not unlikely that mistakes are made in the checking calculation, thus leading to unnecessary work and uncertainty about the accuracy of what has already been done. What is needed is some kind of independent check which involves only very simple working where mistakes are unlikely to occur. Such a check was provided by what came to be known as 'casting out nines', a form of check depending on remainders when numbers are divided by nine. This goes back to India, where it was well known at least from the tenth century onwards, though it is probably much older.

Initially, casting out nines was used as a check only for multiplication and exact division and not for addition and subtraction. This was because, so long as it was the practice to add together only two numbers at a time, checking by using the inverse process was just as simple. The method of checking calculations by using remainders was transmitted from India to the Arab world for multiplication and division only. It was independently extended to addition and subtraction by both Arabs and Hindus once it became the practice to add together several numbers at a time.

The European mathematicians learned of the remainder checks from Arab works around the twelfth century. We find checks by casting out nines, casting out sevens, and casting out elevens, in the writing of Leonardo of Pisa

(Fibonacci). In later European writings and also in some Arab works we find instructions for casting out thirteens and casting out nineteens as well.

Suppose that we multiply 2164 by 462. If we use the lightning method, we shall just have

$$\begin{array}{r} 2164 \\ 462 \\ 999768 \end{array}$$

How can we see quickly if this is likely to be the correct answer? The rule of casting out nines tells us to add the individual digits of each of the numbers which we have multiplied together, to divide each sum by 9, and to multiply the remainders together, again casting out nines if need be. We now add the digits of the product of the two numbers and divide this sum by 9. The remainder obtained from this last operation should equal the product of the two other remainders. This sounds complicated until we try the actual example. We add 2, 1, 6 and 4. This gives us 13, which has remainder 4 on division by 9. Now we add 4, 6, and 2, giving us 12, which has remainder 3 on division by 9. Four times three again gives 12, and remainder 3 on division by 9. Finally, we add 9, 9, 9, 7, 6, and 8 to obtain 48, which also has remainder 3 on division by 9. This agrees with what we obtained before, so the original calculation is probably correct.

Now there are several things which we should notice. First, the remainders 4 and 3 are just the sums of the digits of 13 and 12 respectively. Similarly, the remainder on dividing 48 by 9 can be obtained by adding 4 and 8 to give 12, and then adding 1 and 2. This works generally, that is, remainders on division by 9 can always be obtained simply by summing digits without ever having to divide by 9 at all. Also, we stated that the calculation had been shown to be 'probably' correct. Now interchanging two digits in a number makes no difference to the results of a check. We could, for example, have obtained the incorrect product 999678, where the 6 and the 7 have been interchanged. Because the check depends on adding the digits it makes no difference; such a mistake would not be picked up. Again, had we written one or more of the nines as a zero, the check would not have picked this up either, since any error which changes a sum by 9 or a multiple of 9 will not affect the remainder. This check is therefore by no means infallible, though it is extremely valuable and was therefore very widely used for many centuries.

To see why casting out nines works with decimal calculations we have to look at the remainders when the various powers of ten are divided by 9. We find that 1, 10, 100, 1000, 10 000, and so on, all give remainder 1 when divided by 9. Now any decimal number has the form:

$$a_n 10^n + a_{n-1} 10^{n-1} + \ldots + a_2 10^2 + a_1 10 + a_0$$

where $a_n, a_{n-1}, \ldots, a_2, a_1, a_0$ are all numerals taking values 0 to 9. Division is distributive over addition just as multiplication is distributive over addition – after all, we can just regard division by a number as multiplication by its reciprocal! This means that we can consider each of the places in a decimal number separately. Since each power of ten when divided by 9 has remainder 1, $a_n 10^n$ will have remainder a_n times 1, that is a_n. If a_n equals 9, the remainder is 0. Similarly $a_{n-1}10^{n-1}$ will have remainder a_{n-1}, and so on. The sum of all the remainders will just be the sum of $a_n, a_{n-1}, \ldots, a_2, a_1, a_0$. Now, this may well be equal to more than 8, the maximum remainder on division by 9. In this case, we just repeat the process: digits are summed until a remainder less than 9 is obtained.

It is not difficult to develop an arithmetic of remainders after division by any number. Such remainders may be added, subtracted, multiplied and divided; they are, after all, just numbers. So, if we write $R_n(a)$ to denote the remainder after dividing a by n, we find that

$$R_n(a + b) = R_n(a) + R_n(b)$$
$$R_n(a - b) = R_n(a) - R_n(b)$$
$$R_n(a \times b) = R_n(a) \times R_n(b)$$

Division is not quite as straightforward except when the number to be divided is an exact multiple of the divisor. However, we can still use the check for division, taking into account the main remainder left at the end of the calculation.

Let us look at some further examples with the remainders after division by 9. First, an addition:

$$\begin{array}{rl} 78913 & 28 \rightarrow 10 \rightarrow 1 \\ 4581 & 18 \rightarrow 9 \rightarrow 0 \\ \hline 83494 & 28 \rightarrow 10 \rightarrow 1 \end{array}$$

The remainders on the right imply that the addition is probably correct. Note, however, that the incorrect calculation

$$\begin{array}{rl} 78913 & 28 \rightarrow 10 \rightarrow 1 \\ 4581 & 18 \rightarrow 9 \rightarrow 0 \\ \hline 83404 & 19 \rightarrow 10 \rightarrow 1 \end{array}$$

also appears to be probably correct. Next, subtraction:

$$\begin{array}{rl} 6825 & 21 \rightarrow 3 \\ 3916 & 19 \rightarrow 10 \rightarrow 1 \\ \hline 2909 & 20 \rightarrow 2 \end{array}$$

Here, the remainders agree; 3 minus 1 equals 2. Finally, long division:

```
      166                13 → 4
  38⟌6321        11 → 2⟌12 → 3
     38
     252
     228
      241
      228
       13                  4
```

Here we check by multiplying the divisor and quotient remainders together and adding that obtained from the general remainder, that is, we have 2 times 4 plus 4 equal to 12, which has remainder 3. This should agree with the remainder from the number being divided, which it in fact does.

There are examples of calculations with casting out nines in mathematical documents from many countries. They are to be found especially in works on computation, written in Europe from medieval times onwards, using the new arithmetic associated with the Hindu–Arabic numerals. Most of the methods of calculation discussed earlier are found with casting out nines – particularly multiplications. The lattice method, for example, is often found together with its check. We have, for example,

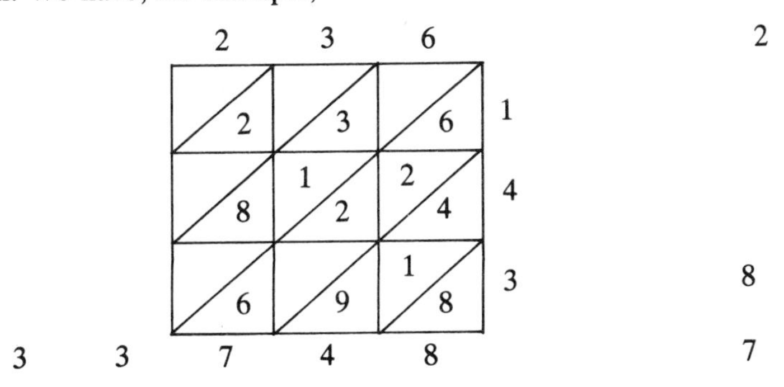

Often, the check was written in the form of a cross, so that for the same multiplication we would find:

16
2 8
7

Here the top position is used for the product of the remainders from the two numbers being multiplied. This has to agree with the remainder from the overall product written in the bottom position. The castle method of multiplying owes its name to the general appearance of a castle given by the overall shape of the calculation itself together with casting out nines. A typical example is

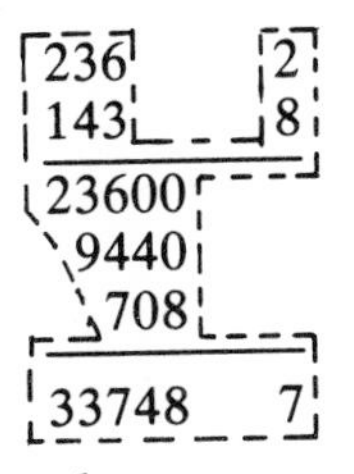

Here we have sketched in the castle.

It was probably because of the element of uncertainty with casting out nines that other checks came to be used. The general principle that division by any number would do for the purposes of checking calculations was known in medieval times, though the formal theory of congruences on which the checks depend for their justification is usually attributed to the great mathematician, Carl Friedrich Gauss, who died in 1855. In fact, much of the theory appears in works by earlier mathematicians, Pascal, Euler and Legendre, for example, as well as others.

The simplest of the other checks is casting out elevens. This involves alternately adding and subtracting the digits of numbers instead of just adding them. To see why this is necessary, we look again at the remainders obtained from the various powers of ten, which we can list as

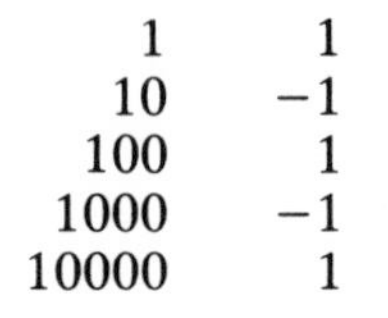

1	1
10	−1
100	1
1000	−1
10000	1

and so on. Here we have taken the remainders which are smallest in magnitude, positive or negative. Thus we can regard 11 as going into 10 no times with remainder 10 or as going once with remainder −1. Either approach is valid, and we have chosen the latter because it illustrates the way in which casting out elevens works. The remainders appear alternately as 1 and −1. So, following a similar principle to casting out nines but taking into account the changes in sign, the remainder on dividing

$$a_n 10^n + a_{n-1} 10^{n-1} + \ldots + a_2 10^2 + a_1 10 + a_0$$

by eleven is just $a_0 - a_1 + a_2 - \ldots$.

We will apply this to the sum of 78913 and 4581 – a calculation which we have already seen.

78913	3 − 1 + 9 − 8 + 7 = 10
4581	1 − 8 + 5 − 4 = −6
83494	4 − 9 + 4 − 3 + 8 = 4

The check, 10 minus 6 equals 4, indicates that according to casting out elevens, the sum obtained is correct. If we apply the check to the incorrect calculation

78913	10
4581	−6
83404	4 − 0 + 4 − 3 + 8 = 13 → 2

we see that casting out elevens, unlike casting out nines, does pick up that we have made an error. Although casting out elevens is not absolutely infallible, it is not fallible in the same ways as casting out nines. Hence, carrying out both checks does virtually guarantee the correctness or otherwise of a computation always assuming, of course, that the checks themselves have been accurately carried out.

With both these checks there are a few ways in which the work involved can be reduced. For example, when casting out nines we can drop nines out as we go along, so that, in working out the remainder for 496821 we can neglect the 9 which follows the 4, then say '4 plus 6 gives 10, remainder 1', and then carry on by saying '1 plus 8 gives remainder 0' and finally '2 plus 1 gives 3'. This achieves the same result as saying '4 plus 9 plus 6 plus 8 plus 2 plus 1 gives 30, which has remainder 3', but with less effort. When casting out elevens, we can ignore any two identical consecutive digits, so that, in finding the remainder on dividing 688774 by 11, we ignore the two 8s and the two 7s and just calculate 4 minus 6 equals −2.

Next in popularity after casting out nines and casting out elevens was casting out sevens. This was normally done by division, but it is possible to devise a shorthand way for obtaining remainders. If we look at the remainders on dividing powers of ten by 7, again taking the smallest remainder irrespective of sign, we have

1	1
10	3
100	2
1000	−1
10000	−3
100000	−2

and the cycle then continues. To check the sum

$$\begin{array}{r} 78913 \\ 4581 \\ \hline 83494 \end{array}$$

we find the remainders on division by 7, starting from the units digit in each case, as follows:

$$[(3 \times 1) + (1 \times 3) + (9 \times 2)] - [(8 \times 1) + (7 \times 3)] = -5 \rightarrow 2$$

the remainder for 78913. For 4581 we have

$$[(1 \times 1) + (8 \times 3) + (5 \times 2)] - (4 \times 1) = 31 \rightarrow 3$$

and for 83494,

$$[(4 \times 1) + (9 \times 3) + (4 \times 2)] - [(3 \times 1) + (8 \times 3)] = 12 \rightarrow 5$$

The check is 2 plus 3 equals 5, indicating that the calculation is correct.

It is possible to work out checks for any divisor and, indeed, some Arab and European medieval works include checks by casting out numbers such as thirteen and nineteen. It seems improbable that these checks were ever widely used, though the modern practice of writing large numbers in groups of three numerals, does make checks by casting out sevens or thirteens viable. This is because 1, 1000, 1 000 000, and all powers of ten of the form 10^{3n} have successive nearest remainders 1, −1, 1, −1, and so on, when divided either by 7 or by 13. So, if we write a large number in the form

$$abc\ def \ldots \ldots \ldots \ldots rst\ uvw\ xyz$$

where $a, b, \ldots, z$ are all digits 0 to 9, we can obtain the remainder on division by either 7 or 13 by alternate addition and subtraction of three-digit numbers in a manner analogous to obtaining the remainder when casting out elevens in the normal way. For example, suppose that we wish to check the multiplication

$$\begin{array}{r} 796582 \\ 84125 \\ 67012460750 \end{array}$$

Marking off each number in threes and alternately adding and subtracting gives us

$$\begin{aligned} 582 - 796 &= -214 \\ 125 - 84 &= 41 \end{aligned}$$

and

$$(750 + 12) - (460 + 67) = 235$$

Casting out sevens from these three numbers, taking account of the minus sign, yields -4, -1, and 4 respectively. The check, -4 times -1 gives 4, indicates that the calculation is correct. Similarly, casting out thirteens gives us -6, 2, and 1 respectively. Now 1 can be taken as -12 because, if a number is one more than a multiple of 13, it is also 12 less than a multiple of 13. The check, -6 times 2 gives -12, confirms the correctness of the original calculation. Clearly, checking in this sort of way is not particularly easy and is likely to require extra calculation on paper. It is therefore not really suitable as a test for normal arithmetic operations. However, it is possible to arrange for such a check to be carried out internally when computation is being performed by an automatic computer.

The value of all these tests which we have been discussing depends essentially on their being easier to carry out and hence less liable to error than what is being tested. It can be argued therefore that it is really only the test by casting out nines that is simple enough, since it involves the easiest of all operations, addition. Alternate addition and subtraction, even with the comparatively simple check by casting out elevens, is liable to lead to computational mistakes in the test itself. It is not at all surprising, therefore, that it was casting out nines that was used most widely over a long period of time even though it cannot guarantee the correctness of any particular calculation.

Ratio and Proportion

Ratios of whole numbers and ratios of lengths were studied and discussed by Greek mathematicians and, indeed, became the focal point for the development of some of the most elegant ideas about number in the ancient world. Here, it is sufficient to note that the Greek philosopher–mathematicians used the basic concept of the ratio of two whole numbers as a means of avoiding having to accept fractions as numbers in their sense of the word.

For us, ordinary fractions such as 1/2, 2/3, 5/19, and so on, are all numbers. We have no difficulty in accepting this because our definition of a number allows for fractions, for irrational numbers such as the square root of two, and indeed for transcendental numbers such as *pi* to be included. For the Greeks, 3/5 was not a number but a ratio between the numbers 3 and 5. For them, only the positive whole numbers were, strictly speaking, numbers at all.

The Pythagoreans in particular were fascinated by certain specific ratios, and especially by those which relate what we call today the arithmetic, geometric, and harmonic means.

The arithmetic mean A of two numbers a and b is just half their sum, that is,

$$A = \frac{a + b}{2}$$

We often refer to this simply as the 'average'.

The geometric mean G of two numbers is the square root of their product, that is,

$$G = \sqrt{ab}$$

The harmonic mean H is less well-known than the other means. It is the reciprocal of the arithmetic mean of the reciprocals. So, if we first take the reciprocals of two numbers a and b, we have $1/a$ and $1/b$. The arithmetic mean of these reciprocals is

$$\frac{\frac{1}{a} + \frac{1}{b}}{2}$$

which simplifies to

$$\frac{a + b}{2ab}$$

The harmonic mean is the reciprocal of this last expression, that is,

$$H = \frac{2ab}{a + b}$$

The particular ratios between these means in which the Pythagoreans were interested were those between the arithmetic and geometric means and between the geometric and harmonic means, and also between the arithmetic and harmonic means and the original numbers. The first two of these ratios were equal, that is, using the colon to denote ratio,

$$A:G = G:H$$

The other relationship is expressed by

$$a:A = H:b$$

The Greeks knew these as the 'perfect' proportion and the 'golden' proportion respectively. They may well have been learned from the Babylonians by Pythagoras himself during the period when he was in Babylon after having been taken prisoner in Egypt.

Ratios lay at the heart of the Pythagorean theory of music. If a string is divided into twelve parts, the ratio 12:6, or 2:1, gives us the octave. If the arithmetic and harmonic means of 12 and 6 are now taken, we have

$$A = \frac{6 + 12}{2} = 9$$

and

$$H = \frac{2 \times 6 \times 12}{6 + 12} = 8$$

The ratios 9:6 and 12:8, both equal to 3:2, correspond to the fifth in the theory of music. Similarly, the ratios 8:6 and 12:9, both equal to 4:3, correspond to the fourth. In this way certain intervals, crucial in the theory of music, were all obtained by ratios involving the numbers 1, 2, 3 and 4, which came to be of mystical significance for they also represented the perfect triangle, yielding the triangular number 10, the sum of 1, 2, 3, and 4.

These ratios for the musical fifth and fourth were used by the Pythagoreans to obtain the whole tone of the diatonic scale,

$$\frac{3}{2} : \frac{4}{3} = 9{:}8$$

and the semi-tone,

$$\frac{4}{3} : \left(\frac{9}{8}\right)^2 = 256{:}243$$

Further mathematical extensions to the theory of music were undertaken by Archytas around 400 B.C. It was however Eudoxus, a pupil of Archytas, who developed the Greek theory of proportions largely as a result of certain philosophical and mathematical difficulties experienced over the discovery of irrational numbers.

Perhaps the use of the idea of proportion most familiar to present-day schoolchildren is its application to problems of the type: 'if 6 apples cost 40 new pence, how much will 9 apples cost?' Although this kind of problem may often be expressed algebraically, it has traditionally been regarded as belonging to arithmetic.

As a problem of arithmetic, proportion has a very long history. The methods of solution have had various names, the most common being 'the golden rule' and 'the rule of three'. The latter can be traced back in India at least to a period before the birth of Christ. It obviously derives from the fact that three quantities have been given and are to be manipulated according to some given rule so as to produce a required fourth quantity. The three given quantities even had special names: they were known as the 'argument', the 'fruit', and the 'requisition'. In the example quoted above, 6 would be the argument, 40 the fruit, and 9 the requisition.

The ancient form of the rule of three requires that the three numbers be set out in such a way that the required answer is obtained by multiplying the second and third numbers and then dividing by the first. For our example, this would mean that the given quantities be written out in the order (in our notation)

$$6 \quad 40 \quad 9$$

The solution is then

$$\frac{9 \times 40}{6} = 60$$

the 60 being in this case new pence.

We can find many examples involving fractions in Hindu documents dating from the seventh century onwards. A typical example is the following one, which occurs in the *Trisatika* (A.D. 750):

> If one-and-one-quarter *pala* of sandlewood are obtained for ten-and-one-half *pana*, how much will nine-and-one-quarter *pala* cost?

The working appears in the form

1	10	9
1	1	1
4	2	4

which can be re-expressed as

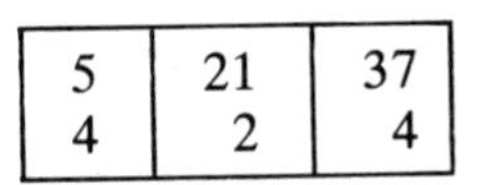

5	21	37
4	2	4

Here, the quantities have been converted from mixed numbers (whole numbers and fractions) to fractions. There is now a further rearrangement,

21	2
37	4
4	5

where the multiples are on the left and the divisors are on the right. This corresponds to our modern way of writing

$$\frac{21 \times 37 \times 4}{2 \times 4 \times 5}$$

The required answer is now obtained as 77 7/10 *pana*.

The Indian mathematicians were equally familiar with the inverse rule of three where the proportion is reversed. The corresponding rule required that the middle and first terms be multiplied and then divided by the third term.

Thus, we follow the same method of writing down the numbers as before, when we solve the problem:

> If three men build a wall in 5 days, how many days will it take 4 men to build the same wall, working at the same rate?

We write down

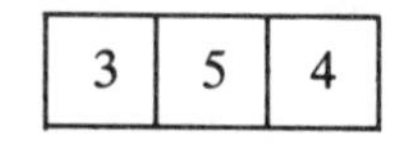

and then, multiplying five by three and dividing by four, we obtain the answer, 3 3/4 days.

The principles of the rule of three were extended to problems involving 5, 7, 9, and even 11 terms. In all these cases, the basic idea was the same, the result being obtained by multiplying and dividing by appropriate terms which were first collected together on the multiplier and divisor sides of a lattice.

The rule of three was transmitted from India to the Arabs in the eighth century, and was subsequently passed on to western Europe. The original Hindu name was usually retained, though it also had other names. It is described in detail by a number of famous mathematicians such as al-Khowarizmi, Leonardo of Pisa, Roberte Recorde, Nicolas Chuquet, and many others. It found its way especially into commercial arithmetics. Often, its application was extended to dividing a given amount into unequal parts according to a fixed proportion. Chuquet, for example, discusses how 100 may be divided into two parts with ratios 7 to 9 simply by application of the rule of three. First, we must add 7 and 9 to obtain the common divisor, 16. We now just apply the rule twice.

16	100	7

giving

$$\frac{100 \times 7}{16} = 43\tfrac{3}{4}$$

and

16	100	9

giving

$$\frac{100 \times 9}{16} = 56\tfrac{1}{4}$$

In fact, Chuquet sets out the calculation a little differently, writing

$$100 \begin{cases} 7 & 43\tfrac{3}{4} \\ 9 & 56\tfrac{1}{4} \end{cases}$$
$$\overline{16}$$

but he specifically calls this the rule of three.

Prime Numbers and Factors

The highest common factor (h.c.f.; or the greatest common divisor as it has often been called) and the lowest common multiple (l.c.m.) of two or more numbers can most readily be found by decomposing the numbers into their prime factors. For example, the h.c.f. and l.c.m. of 3900, 2520 and 1680 follow immediately from the decompositions of these numbers, which we may write as follows:

$$\begin{aligned} 3900 &= 2 \,.\, 2 \,.\, 3 \,.\, 5 \,.\, 5 \,.\, 13 \\ 2520 &= 2 \,.\, 2 \,.\, 3 \,.\, 5 \,. \qquad\quad 2 \,.\, 3 \,.\, 7 \\ 1680 &= 2 \,.\, 2 \,.\, 3 \,.\, 5 \,. \qquad\quad 2 \,. \quad\; 7 \,.\; 2 \end{aligned}$$

The h.c.f. is by definition the largest whole number which will divide exactly into each of the given numbers. It follows therefore that it is the product of all those factors which are common to each. We can therefore write down immediately:

$$\text{h.c.f.} = 2 \,.\, 2 \,.\, 3 \,.\, 5 = 60$$

The l.c.m. is by definition the smallest whole number into which each of the given numbers will divide exactly. It follows therefore that it is the product of all those factors which occur in one at least of the given numbers. We can therefore write down immediately:

$$\text{l.c.m.} = 2 \,.\, 2 \,.\, 3 \,.\, 5 \,.\, 5 \,.\, 13 \,.\, 2 \,.\, 3 \,.\, 7 \,.\, 2 = 327\ 600$$

Factors which are repeated, such as 2, 3, and 5, have to be taken into account the appropriate number of times. The validity of this method depends on the fact that any whole number can be uniquely decomposed into its prime factors.

Prime numbers have fascinated mathematicians from Greek times onwards. In Book VII of Euclid's *Elements* we find the first known definition of a prime number:

> a prime number is that which is measured by the unit alone.

By contrast:

> a composite number is that which is measured by some number.

Thus, all other whole numbers are products of prime numbers. The primes can be regarded as the bricks from which all other whole numbers are obtained by multiplication.

One of the earliest known methods for finding prime numbers is that known as the sieve of Eratosthenes. If we write down all the integers as far as we want to go, we can then strike out all those divisible by 2, divisible by 3, divisible by 5, and so on in turn, and we shall eventually be left with the prime numbers. In fact, we need only proceed until we are dividing by the largest prime whose square is less than the greatest number we are considering. Thus, if we are looking for all the prime numbers up to 100, we need to sieve out only numbers divisible by 2, 3, 5 and 7. The square of the next prime number, 11, is greater than 100.

1	2	3	~~4~~	5	~~6~~	7	~~8~~	~~9~~	~~10~~
11	~~12~~	13	~~14~~	~~15~~	~~16~~	17	~~18~~	19	~~20~~
~~21~~	~~22~~	23	~~24~~	~~25~~	~~26~~	~~27~~	~~28~~	29	~~30~~
31	~~32~~	~~33~~	~~34~~	~~35~~	~~36~~	37	~~38~~	~~39~~	~~40~~
41	~~42~~	43	~~44~~	~~45~~	~~46~~	47	~~48~~	~~49~~	~~50~~
~~51~~	~~52~~	53	~~54~~	~~55~~	~~56~~	~~57~~	~~58~~	59	~~60~~
61	~~62~~	~~63~~	~~64~~	~~65~~	~~66~~	67	~~68~~	~~69~~	~~70~~
71	~~72~~	73	~~74~~	~~75~~	~~76~~	~~77~~	~~78~~	79	~~80~~
~~81~~	~~82~~	83	~~84~~	~~85~~	~~86~~	~~87~~	~~88~~	89	~~90~~
~~91~~	~~92~~	~~93~~	~~94~~	~~95~~	~~96~~	97	~~98~~	~~99~~	~~100~~

Sieve of Eratosthenes

However, mathematicians through the ages have been interested in the actual distribution of the prime numbers and in finding formulae from which they can be obtained. We know that there is an infinite number of primes, Euclid's proof of this having long been held as a model of mathematical proof structure. Repeated attempts to find a formula from which all prime numbers can be obtained have, however, failed. One of the most famous of these was that of the French seventeenth-century mathematician, Pierre de Fermat, who announced that all numbers of the form

$$(2)^{2^n} + 1$$

are prime. In fact, this is true for n equal to 0, 1, 2, 3, and 4, but it is not true for n equal to 5. This was discovered a century later by Leonhard Euler.

Another famous conjecture about prime numbers is that of the eighteenth-century mathematician, Christian Goldbach. He stated, without proof, that every even number is the sum of two prime numbers, and that every odd number is either prime or the sum of three prime numbers. The second part of this, in fact, follows from the first. The first part has never been proved. It remains one of many conjectures in number theory (the study of properties of whole numbers) which seem to be true, but which have not so far been proved.

We are not concerned here, however, with the history of number theory, but with the factorization of composite numbers into their constituent primes. Tests on numbers to find their prime factors have been known for many centuries. Clearly, all even numbers and only even numbers are divisible by two. We have already seen that remainders on division by 9 can be obtained by summing the individual digits of numbers. Thus, if the process of repeated summing of digits yields 3, 6, or 9, the number concerned must be divisible by 3. Any number ending in 5 or 0 is divisible by 5.

The check by casting out elevens is equally linked to a corresponding divisibility test. We simply add and subtract the individual digits of a number. For example, we establish that 985182 is divisible by 11 by the calculation

$$9 - 8 + 5 - 1 + 8 - 2 = 11$$

Similarly, to test the same number for divisibility by seven, we would calculate

$$(1 \times 2) + (3 \times 8) + (2 \times 1) - (1 \times 5) - (3 \times 8) - (2 \times 9) = -19$$

showing that it is not divisible by 7. Tests of this type can be similarly determined for divisibility by any prime number.

An alternative method of finding the h.c.f. of any two numbers is provided by Euclid's algorithm. This does not require us to find all the factors of the numbers. It requires us to subtract the smaller of the two numbers from the larger as often as possible, then to subtract the remainder from the smaller number as often as possible, to subtract the new remainder from the previous one, and so on. The last non-zero remainder is the h.c.f.

Clearly, we can obtain the remainders by division more economically than by repeated subtraction, and we do this for two of the numbers in the examples with which we began:

$$\begin{aligned} 3900 &= 2520 + 1380 \\ 2520 &= 1380 + 1140 \\ 1380 &= 1140 + 240 \\ 1140 &= (4 \times 240) + 180 \\ 240 &= 180 + 60 \\ 180 &= (3 \times 60) + 0 \end{aligned}$$

The required h.c.f. is therefore 60, the last non-vanishing remainder.

To understand why Euclid's algorithm works, let us suppose that a and b are the two numbers, and that a is greater than b. In general algebraic terms, we have

$$a = q_1 b + r_1$$

where q_1 is the remainder. We continue:

$$\begin{aligned} b &= q_2 r_1 + r_2 \\ r_1 &= q_3 r_2 + r_3 \\ &\cdots \quad \cdots \quad \cdots \\ r_{n-2} &= q_n r_{n-1} + r_n \\ r_{n-1} &= q_{n+1} r_n + 0 \end{aligned}$$

Each remainder must be smaller than the preceding one as the sequence of divisors is decreasing, so we must eventually arrive at a zero remainder. Now, since r_n divides r_{n-1}, it must also divide r_{n-2} since it divides both the terms $q_n r_{n-1}$ and r_n. This argument carried upwards, proves that r_n is a common factor of a and b. We still have to show, however, that it is the highest of the common factors of a and b. This will follow if we can show that all other common factors divide exactly into it. Suppose, therefore, that c is any common factor. This must exactly divide r_1, because it exactly divides a and $q_1 b$. It must also exactly divide r_2 because it exactly divides b and $q_2 r_1$. Carrying the argument downwards, we have eventually that it must exactly divide r_n.

Mathematicians are usually interested in the most efficient way of performing calculations. Clearly, Euclid's algorithm can involve a considerable number of calculations. It was eventually proved that the number of divisions involved is at most five times the number of digits of the smaller number. Later, it transpired that the most efficient method of carrying out the procedure is to use least absolute remainders, that is, to take the nearest multiple at each stage irrespective of whether it is greater or less than the number being divided. For our example, we would then have

$$\begin{aligned} 3900 &= (2 \times 2520) - 1140 \\ 2520 &= (2 \times 1140) + 240 \\ 1140 &= (5 \times 240) - 60 \\ 240 &= (4 \times 60) + 0 \end{aligned}$$

In this case, by using least absolute remainders we have reduced the divisions from six to four.

The h.c.f. and the l.c.m. of two numbers must be related to each other. By looking at our first calculation of the h.c.f. and l.c.m. (page 127) we can see that the l.c.m. of two numbers must be their product divided by their h.c.f.

One advantage of the factorization method is that we can very readily obtain the h.c.f. and l.c.m. of several numbers at one and the same time. If we wish to use Euclid's algorithm for more than two numbers, we have to use it more than once. Thus, for our original three numbers, we first use it to find that the h.c.f. of 3900 and 2520 is 60, and then use it again to find that the h.c.f. of 60 and 1680 is also 60.

The importance of the h.c.f. and l.c.m. lies in their application to fractions. It was the practical use of fractions in commercial calculations which led to the discovery of methods for calculating them. In order to reduce a fraction to its lowest terms, we divide both numerator and denominator by their h.c.f. In order to add and subtract fractions, we find the l.c.m. of their respective denominators.

Euclid's algorithm also has important applications in the theory of Diophantine equations, that is, indeterminate equations where we are looking only for whole-number solutions. It enables us to decide for any given equation whether or not a solution actually exists and, if it does, it can also indicate the most direct method of finding that solution. These equations were named after the third-century Greek mathematician, Diophantus of Alexandria.

The Calculation of Roots

Methods, which can provide schemes for calculating roots of any order, have been known for many centuries. We shall restrict ourselves here to looking at the calculation of square and cube roots.

In ancient times recourse was made to tables of roots. Thus, for example, the Babylonian mathematicians looked up both square and cube roots in tables whenever they were needed in the course of their calculations. They were not concerned with the problem of the irrationality of roots as they were able to obtain approximate values sufficiently accurate for any purpose to hand. A surviving tablet gives as an approximation for the square root of two the sexagesimal number 1 24 51 10, that is:

$$1 + \frac{24}{60} + \frac{51}{3600} + \frac{10}{216\,000}$$

We do not know just how the Babylonians calculated their roots with such accuracy, though we do know that at least one method for obtaining square roots to any required degree of accuracy was known to both Greek and Hindu scholars, and it is a reasonable hypothesis that the Greeks may have obtained their knowledge from the Babylonians. We shall begin, however, by considering roots of numbers which are perfect squares.

We find descriptions of methods for calculating square roots in many early Indian writings. The principal method is very similar in many respects to that taught in schools today. We can follow its workings by looking at a specific example, the square root of 59 049.

First of all, the number was written down and the odd and even places marked with small vertical and horizontal strokes respectively:

$$\begin{array}{c} | - | - | \\ 59049 \end{array}$$

The calculation now proceeded:

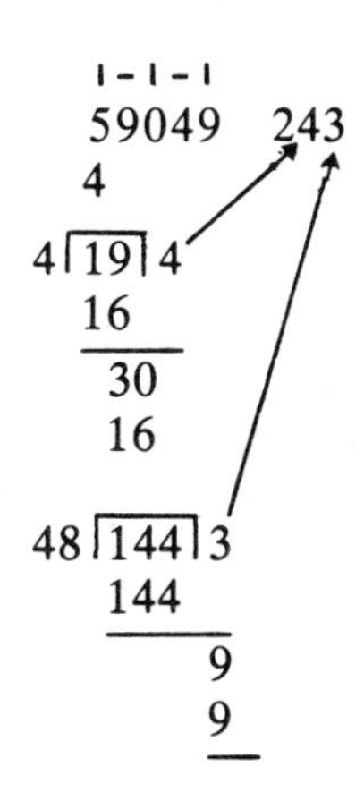

Subtract square next below the number as far as the first odd place and write down the root.

Divide up to next even place by twice the root and put quotient on root line;

subtract.

Subtract square of quotient from number up to next odd place.

Divide number up to next even place by twice root and put quotient on root line:

subtract.

Subtract square of last quotient from number up to next (the last) odd place.

The process ends and the required root is 243. Of course, we have used modern forms of numerals in the example and set out the calculation in a way which has made the explanation of the stages meaningful. Basically, there are two stages, repeated alternately. For the odd places we subtract squares, and for the even places we divide by twice the root already calculated.

The method is, however, not always quite as straightforward as the example might suggest. We will calculate the square root of 66 564, using exactly the same process.

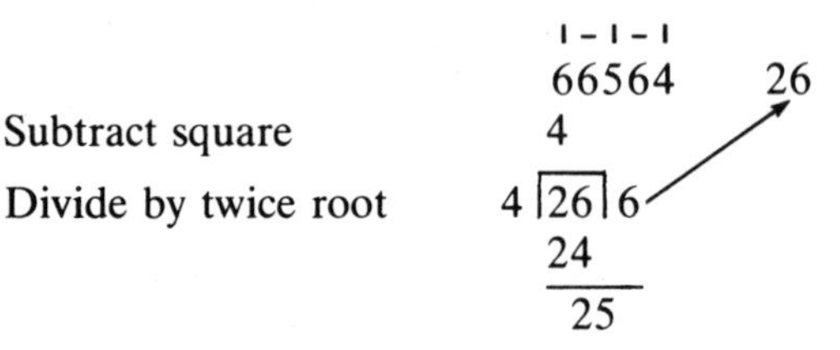

We should now subtract the square of the quotient 6, namely 36, from 25. We cannot do this, however, so we have to go back a stage and reduce the quotient 6 to 5.

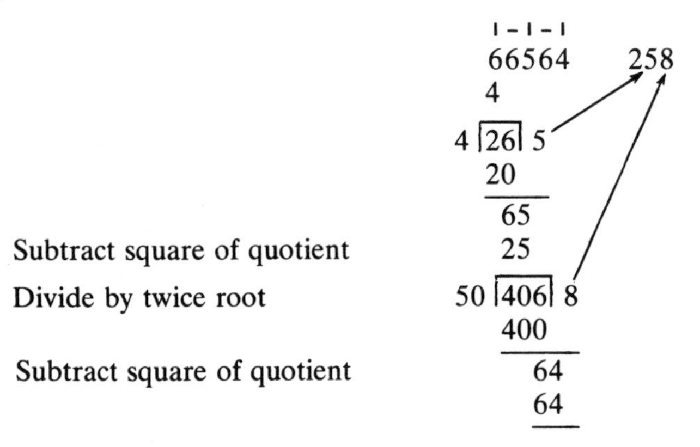

We have now obtained the correct root, 258.

Making a correction in the calculation, such as the change of quotient from 6 to 5, was a perfectly simple operation on a dust board; the numeral 6 was merely rubbed out together with the unwanted part of the calculation associated with it, 5 was substituted and the calculation was then continued in the usual way. At the end of the calculation there would be no trace of the adjustment. In later written calculations, it became a matter of crossing out digits and writing the new digits either above or below, or sometimes alongside the digits crossed out. In our modern method, bringing down digits in pairs avoids the need for making such corrections.

To see why this ancient method works we need to think about what happens when we square a number. Suppose that we start with a three-digit number. We can represent this in the general form

$$100a + 10b + c$$

where a, b and c are the three digits which we write down – in our earlier example these would be 2, 4 and 3. Squaring the general expression gives us:

$$10000a^2 + 2000ab + 100b^2 + 200ac + 20bc + c^2$$

a number made up from the sum of six parts and corresponding in our first example to 59 049. In practice, if we choose a sufficiently large three-digit number to start with, we shall find that its square has 6 digits. It cannot, of course, have more digits than 6 since the square of 999 is 998 001. In order to obtain the square root, we need to subtract the various parts successively in such a way that a, b and c, the digits of the root, will be clearly indicated. If we now follow through in general terms exactly what we did when we calculated the root of 50 049, we shall see just how the method works.

When we subtract the square next below the number taken as far as the first odd digit from the left, we are subtracting $10\,000a^2$, in the example 40 000. Note that it is not just a^2 that we subtract, but a^2 in the fifth place from the right. The next subtraction is $2000ab$, that is twice the initial root a times the quotient b all taken four places from the right, in our example 16 000. The next subtraction is just $100b^2$, that is b^2 taken three places from the right, in our example 1600. We have now a number representing

$$200ac + 20bc + c^2$$

still left. The next subtraction is $(200a + 20b) \times c$, that is, twice the root $10a + b$ (obtained so far) times the quotient c, all taken two places from the right. In our example, this corresponds to 1440. Finally, we subtract c^2, in our example 9, which should and does indeed subtract exactly.

This method of extracting square roots was transmitted to the Arabs, and examples of its application by them can be traced back to the ninth century. In parallel with the division methods which we discussed earlier, there was a galley or scratch method of setting out calculations of square roots. Here, our calculation of the square root of 59 049 would look like

```
 1
134
59049     243
.  .  .
24483
  4
```

Note that dots were first placed underneath alternate digits so that the correct starting position was made clear, and one of the factors was placed beneath an appropriate dot each time a square was subtracted. Otherwise spaces were filled in the same way as for the division on page 114.

As the scratch method of division gradually gave way from the fifteenth century onwards to the modern method, so the scratch method of setting out the calculation of square roots gave way to the modern method where the calculation looks indeed very much like long division. As with division, there were slight variations, but the main contrast with earlier method is the combination of the subtraction processes in pairs so that two digits of the original number are taken at a time. The modern form of the calculation of the square root of 66 564 thus appears as

```
       2 5 8
     ________
   2 |66564
      4
     ____
  45 |265
      225
     _____
 508 |4064
      4064
```

though sometimes the root is placed on the right rather than at the top of the calculation. Here, for example, just before the second subtraction we arrive at this stage:

```
       2
     ______
   2 |66564
      4
     ____
   4 |265
```

Having doubled the root, 2, we leave a space to the right of the 4 and decide on the largest number which we can insert and use also to multiply the divisor thus obtained without exceeding 265. This is, of course, 5. Two operations of the older method are thus combined and there is no problem of being initially misled into choosing 6 rather than 5 and having subsequently to retrace our steps.

This principle of square root extraction is not confined to finding the roots of exact squares provided that a place-value system of numerals is available. It was used to obtain roots to any required number of places using the sexagesimal notation of the Babylonians. The earliest Arab example is a sexagesimal calculation. With the advent of decimal fractions in the West, both division and square root extraction could be extended to fractional places simply by including the decimal point and the required number of zeros to the right of the original number. This is just what we do today.

There is, however, a method of square root extraction, also known from ancient times, which relies on entirely different principles. It was certainly known to the Greeks and the Hindus, and it had diffused to western Europe via the Arabs by about the thirteenth century. As an example we find the square root of 720, which we can lay out schematically as follows:

Write down the number	720
Take next perfect square above	729
Take the square root	27
Divide number by this root	$720 \div 27 = 26\frac{2}{3}$
Add this quotient to the root	$26\frac{2}{3} + 27 = 53\frac{2}{3}$
Divide by two	$53\frac{2}{3} \div 2 = 26\frac{5}{6}$

The result so far is reasonably accurate, as we can tell by squaring it. This gives us 720 1/36, differing from our original number by only 1/36. We can get an even more accurate result by taking 26 5/6 as the root of the nearest square greater than 720, and repeating the process.

This particular example is described in Heron's *Metrica*, dating from the first century. The principle can be expressed in general terms quite simply. Let X be the number whose root we wish to find, and let x_1^2 be the next perfect square greater than X. Taking x_1 as our first approximation to the desired root, the next is taken as

$$x_2 = \tfrac{1}{2}\left(x_1 + \frac{X}{x_1}\right)$$

and the next as:

$$x_3 = \tfrac{1}{2}\left(x_2 + \frac{X}{x_2}\right)$$

and so on. The results successively obtained will exceed the true root by ever smaller amounts, and we can see that this is so in the following way. Assume that the first approximation x_1 exceeds the true root x by an amount a, that is,

$$x = x_1 + a$$

where a will be less than one. We now have for the difference between the true root and the second approximation

$$\begin{aligned} x_2 - x &= \tfrac{1}{2}\left(x_1 + \frac{X}{x_1}\right) - x \\ &= \tfrac{1}{2}\left[x_1 + \frac{(x_1 - a)^2}{x_1}\right] - (x_1 - a) \\ &= \frac{a^2}{2x_1} \end{aligned}$$

which is certainly less than the previous difference a, since a is less than one.

A repeatable approximation method such as this is called an 'iterative' method, and methods of this type form the basis of programmes for computer calculations. The method is clearly different in principle from our more familiar method taught in schools, discussed earlier.

Several other iterative methods for square roots were known and used by the Arabs and by European mathematicians of the Renaissance period and after. Sometimes the calculations led to approximations which were less than the true root, as, for example, the method used by Al-Karkhi and later by the sixteenth-century Italian mathematician Tartaglia, which has the general expression,

$$x_2 = x_1 + \frac{X - x_1^2}{2x_1 + 1}$$

and gives 26 44/53 as the square root of 720, a result which is less accurate than that given by Heron's method. Whatever method was used, however, the means of checking the result obtained was always to square the calculated root and see how closely the original number was approximated. In the case of an exact root, the original number would of course be recovered exactly.

Methods of extracting cube roots date back also to ancient times. Heron sets out a calculation of the cube root of 100. We can reproduce Heron's arrangement of the calculation exactly, but for our discussion here we will translate it into modern numerals. The calculation is as follows:

Nearest perfect cubes above and below number	$125 = 5^3$ $64 = 4^3$
Differences from original number	$125 - 100 = 25$ $100 - 64 = 36$
Larger root times difference from smaller cube	$5 \times 36 = 180$
Smaller root times difference from large cube	$4 \times 25 = 100$
Add	$180 + 100 = 280$
Product of larger root with difference from smaller cube divided by this sum	$180 \div 280 = \frac{9}{14}$
Add to smaller root	$4\frac{9}{14}$

In modern decimal notation this is equal to 4·6428 . . . , which is by no means a poor approximation to the true cube root 4·6416 . . . .

To see how this method works, we must first begin to follow through this calculated example, noting each step in general terms.

Suppose that we want to find the cube root of some number X – in our example, 100. We first find x_1^3 and $y_1^3 = (x_1 - 1)^3$ such that x_1^3 is the nearest perfect cube greater than x and y_1^3 the nearest perfect cube less than X. We assume here, of course, that X is not itself a perfect cube. In our example:

$$x_1^3 = 125$$
$$y_1^3 = 64$$

giving

$$x_1 = 5$$
$$y_1 = 4$$

We now find $x_1^3 - X$ and $X - y_1^3$. We have, in our example.

$$x_1^3 - X = 125 - 100 = 25$$
$$X - y_1^3 = 100 - 64 = 36$$

We next find $x_1(X - y_1^3)$ and $y_1(x_1^3 - X)$ and add; that is, we find

$$x_1(X - y_1^3) + y_1(x_1^3 - X)$$

in our example:

$$(5 \times 36) + (4 \times 25) = 180 + 100 = 280$$

Finally, we divide $x_1(X - y_1^3)$, in our example 5×36, namely 180, by this sum; that is, we calculate:

$$\frac{x_1(X - y_1^3)}{x_1(X - y_1^3) + y_1(x_1^3 - X)}$$

and then we add this to the smaller root y_1 to get the required cube root:

$$x = y_1 + \frac{x_1(X - y_1^3)}{x_1(X - y_1^3) + y_1(x_1^3 - X)}$$

However, we know that this is an approximation and not the exact root. To see what has been discarded in making the approximation we have to go back and write x_1 as $x + a$ and y_1 as $x - b$, where $a + b$, the difference between x_1 and y_1, we know to be unity. If we now cube the smaller of these roots we obtain

$$y_1^3 = (x - b)^3 = x^3 - 3bx^2 + 3b^2x - b^3$$

which is approximately equal to

$$X - 3bx^2 + 3b^2x$$

if we write x^3 as X and neglect b^3. This is not an unreasonable approximation to make since b is less than one and b^3 consequently even smaller. We can now slightly rearrange our expression so that we have

$$X - y_1^3 \simeq 3bx(x - b) = 3bxy_1$$

(Note that we are using the symbol $\simeq$ to denote approximation.) In the same way we can take the cube of the larger root to give us

$$x_1^3 = (x + a)^3 = x^3 + 3ax^2 + 3a^2x + a^3$$

and, neglecting a^3 since a is smaller than one, we get

$$x_1^3 \simeq X + 3ax^2 + 3a^2x$$

and then

$$x_1^3 - X \simeq 3ax(x + a) = 3axx_1$$

If we now divide the two expressions we have

$$\frac{3bxy_1}{3axx_1} \simeq \frac{X - y_1^3}{x_1^3 - X}$$

that is,

$$\frac{b}{a} \simeq \frac{x_1(X - y_1^3)}{y_1(x_1^3 - X)}$$

Now, since the sum of a and b is unity, we can rewrite b as $\frac{b}{b + a}$, and this will be approximately equal to

$$\frac{x_1(X - y_1^3)}{x_1(X - y_1^3) + y_1(x_1^3 - X)}$$

So, the approximate expression for the cube root $x = y_1 + b$ is

$$y_1 + \frac{x_1(X - y_1^3)}{x_1(X - y_1^3) + y_1(x_1^3 - X)}$$

which is exactly the expression which we obtained when we followed through in general terms Heron's calculation of the cube root of 100. It seems not unreasonable to suggest that it was by just such a method that the Babylonians computed their tables of cube roots.

There is also a method of finding cube roots analogous to our familiar modern method for square roots and going back in India to at least the sixth century. We will again look at a specific example, finding the cube root of 1 906 624, calculating according to the instructions found in the *Aryabhatiya*.

Step	Calculation	Root
Group digits into threes	1906624 (grouped as 1 | 906 | 624)	124
Subtract largest possible cube from first group on the left	1	
Divide by three times the square of the root	3 ⟌ 9 2	(quotient 2)
	6	
	30	
Subtract square of quotient times three times previous root	12	
	186	
Subtract cube of quotient	8	
Divide by three times square of root obtained so far	432 ⟌ 1786 4	(quotient 4)
	1728	
	582	
Subtract square of quotient times three times previous root	576	
	64	
Subtract cube of quotient	64	
	—	

Here, we have to perform operations in cycles of three. First, we subtract a cube. Next, we divide by three times the square of the root already calculated. Note that in this example we take the quotient to be 2 and not 3 in order to be able to proceed with the rest of the operations. Thirdly, we subtract three times the root times the square of the quotient just obtained. If we expand in general terms the operation of cubing the three-digit number $100a + 10b + c$, we can see by analogy with our analysis of the corresponding square root method why this particular repeated sequence of three operations works. We shall not do all of this in detail here, but merely point out that the first three terms of the expansion will be

$$10^6a^3 + 3.10^5a^2b + 3.10^4ab^2 + \ldots$$

The link with the cycle of three operations can now be seen. The first three subtractions involve the removal one by one of quantities exactly corresponding to these three terms. The unit digit of the root is exposed by the next subtraction but one, where we effectively subtract $3.10^2(10a + b)^2c$.

By analogy with square root calculations, it later became the custom to deal with the digits of the original number in groups of three digits at a time and to set out the calculation in something akin to division format. The calculation of the cube root of 1 906 624 might therefore in the sixteenth century have looked something like

```
1| 906 | 624 |124                600
1|      |     |                   120
--+------+-----+                     8
  | 906 | 624 |                  ---
  | 3̸0̸0̸ |     |                  728
  | 728 |     |
--+------+-----+
  | 178 | 624 |                172800
  |  4̸3̸ | 2̸   |                  5760
  | 178 | 624 |                    64
                               ------
                               178624
```

This is the way in which, for example, the French mathematician Jean Trenchant presented the calculation of cube roots. Here we see the divisors written in the calculation but then crossed out so that there could be no confusion when subsequently subtracting. The intermediate stages of the calculation are shown on the right. We can easily see that this is just the method of the *Aryabhatiya*. In modern times we would set out the calculation as

```
   1  2  4
1 ⟌1906624
   1
   ----
    906
    728
    ------
    178624
    178624
    ------
```

although the root might be written to the right instead of at the top. However, unless we were already familiar with the principles involved we would find it difficult to see just what were the intermediate stages involved.

As with square roots, the checking of cube root calculations was always carried out by cubing the root obtained and comparing the result with the original number.

Logarithms

Today, we usually approach logarithms through indices. The Greeks had a notation for indices which showed clearly their understanding that, for example, the cube of a number multiplied by the square of the same number yielded its fifth power. We can represent this generally by

$$a^m \times a^n = a^{m+n}$$

The important point here is that multiplication is associated with addition of indices.

This association between multiplication and addition or subtraction became familiar to Renaissance astronomers in the West, who made use of trigonometric identities of the type:

$$\sin A \times \sin B = \tfrac{1}{2}[\cos(A - B) - \cos(A + B)]$$

In this case, multiplication of two trigonometric quantities, $\sin A$ and $\sin B$, is associated with subtraction of the trigonometric quantities $\cos(A - B)$ and $\cos(A + B)$. Although such associations were comparatively well-known in the mathematical world of the fifteenth and sixteenth centuries, the basis from which logarithms were eventually to spring was the comparison of arithmetic and geometric progressions.

In an arithmetic progression each term differs from the immediately following and previous terms by a fixed quantity. The most obvious example is provided by the whole numbers where the fixed difference is one. In a geometric progression each term differs from the immediately following and previous terms in a fixed ratio. An obvious example is provided by the powers of ten which form the basis of our decimal numeral system. So, if we consider a term a somewhere within a progression, for an arithmetic progression we will have

$$\ldots, a-3d, a-2d, a-d, a, a+d, a+2d, a+3d, \ldots$$

where d is the constant difference, and for a geometric progression we will have

$$\ldots, \frac{a}{r^3}, \frac{a}{r^2}, \frac{a}{r}, a, ar, ar^2, ar^3, \ldots$$

where r is the fixed ratio. Thus, for the whole numbers, d equals one, and for the successive powers of ten, r equals ten.

We find a comparison between these two kinds of progression in the fifteenth century *Triparty* of Nicolas Chuquet. He may have been aware of the principle underlying logarithms, though he did not develop a detailed discussion of it. In the middle of the sixteenth century the German mathematician Michael Stifel wrote in considerable detail about the comparison between arithmetic and geometric progressions in a work entitled *Arithmetica integra*, in which we find particular attention drawn to the parallel between the respective additive and multiplicative characteristics.

If we write out a geometric progression with unity as the reference term underneath the whole numbers with zero as the reference term, we can easily see exactly how multiplication and division amongst the terms of the geometric progression correspond to addition and subtraction of terms in the arithmetic progression:

$$\begin{matrix} \ldots & -4 & -3 & -2 & -1 & 0 & 1 & 2 & 3 & 4 & \ldots \\ \ldots & \frac{1}{r^4} & \frac{1}{r^3} & \frac{1}{r^2} & \frac{1}{r} & 1 & r & r^2 & r^3 & r^4 & \ldots \end{matrix}$$

For example, if we multiply $\frac{1}{r^2}$ by r^3, we just get r. Correspondingly, if we add -2 and 3, the numbers above $\frac{1}{r^2}$ and r^3, we just get 1, the number above r.

Stifel went further than this. He noted that taking powers and roots of terms in the geometric progression corresponds to multiplication and division of terms in the arithmetic progression. For example, if we take the cube of r^2, giving us r^6, this will lie under the number which we get when we multiply two by three. Similarly, if we take the square root of r^8, giving us r^4, this corresponds to dividing 8 by 2 to get 4.

In 1614 the Scottish mathematician John Napier published in Latin a work whose title translates as *A Description of the Wonderful Law of Logarithms*. Napier's logarithms were obtained on the principle of associating an increasing arithmetic progression with a decreasing geometric one. However, he saw the relationship in terms of moving points. We can understand this by considering two equal lines AB and CD and supposing that points P and Q move respectively along these lines from A and C, each starting at the same speed.

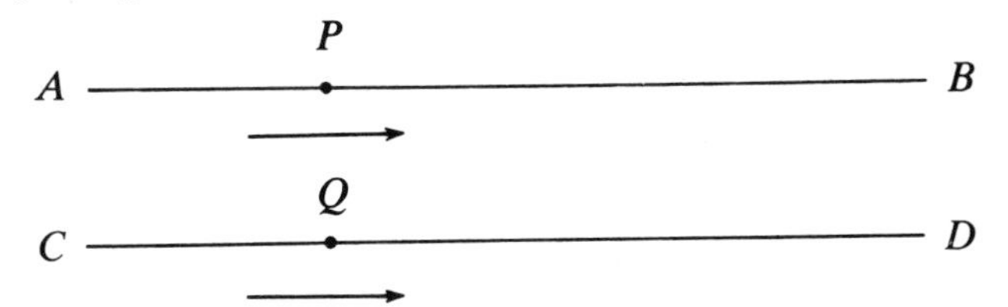

P moves throughout with this constant speed which is numerically equal to the length of AB and CD; Q moves with a decreasing speed always numerically equal to QD. At any given moment, AP is defined to be the logarithm of QD.

Strictly speaking, we have here a situation where there is a continuous change of speed, and we might think that the techniques of the calculus are needed. However, if we consider small enough constant intervals of time and assume that during each interval the speed of Q is constant and equal to its velocity at the beginning of the interval, then successive lengths of QD, corresponding to the beginnings of these intervals will be in geometric progression. Since P travels with constant speed, the corresponding successive lengths AP will be in arithmetic progression.

We can obtain an indication of the way in which Napier calculated his logarithms by looking at the extract from his table below. Here, we start with zero in the arithmetic progression, taking differences of 1, and with

10 000 000 in the geometric progression and a common ratio of

$$1 - \frac{1}{10^7}$$

This common ratio means that we subtract the ten-millionth part from each term in order to obtain the next one. Napier chose this very large number as the first term of his geometric progression in order to avoid fractions, the terms actually representing the sines of angles. The best table of sines then available to him extended to seven places.

Arithmetic progression	*Geometric progression*
0	10 000 000
1	9 999 999
2	9 999 998
3	9 999 997
4	9 999 996
...	...

In terms of Napier's geometrical understanding of logarithms, the assumptions made in constructing this table amount to taking intervals of time such that *P* and *Q* start moving at the speed of one unit of distance in one unit of time, the total length of *CD* being unity, though scaled up by a factor of 10^7 in order to avoid fractions.

Now at first sight, it might appear that the geometric progression is just another arithmetic one with a common difference of unity. This appearance is a coincidence, however, arising from approximating to seven places. The terms are in fact calculated as the geometric progression,

$$10^7, 10^7\left(1 - \frac{1}{10^7}\right), 10^7\left(1 - \frac{1}{10^7}\right)^2, 10^7\left(1 - \frac{1}{10^7}\right)^3, \ldots$$

After a hundred steps, we have arrived at the pair of numbers

100	9 999 999

Here, Napier noticed that the term of the geometric progression could have been obtained directly from 10 000 000 by subtracting one-hundred-thousandth part of it, that is,

$$10^7 - 10^2 = 9\,999\,900$$

He therefore now continued with differences of 100 at a time in the arithmetic progression until he reached the pair of numbers

5000	9 995 001

Here a slight adjustment is made. By adding a further $1\frac{1}{4}$ to 5000, the corresponding term of the geometric progression becomes equal to the original

10 000 000 less one two-thousandth part of it:

5001·25	9 995 000

Napier then proceeded with larger and larger steps until he arrived at a value in the geometric progression a little short of 5 000 000, and at this point he stopped.

These logarithms of John Napier are not the same as natural logarithms, often incorrectly referred to as Napierian logarithms. The latter are logarithms to the base e, a transcendental number whose value is 2·718 282 818 284 . . . , the sum of the infinite series

$$1 + \frac{1}{1} + \frac{1}{2\,.\,1} + \frac{1}{3\,.\,2\,.\,1} + \frac{1}{4\,.\,3\,.\,2\,.\,1} + \cdots$$

Napier's logarithms are defined in terms of the relation between lengths or distances and not in terms of the powers of a given base.

Independently of Napier, a Swiss watch and instrument maker, Joost Bürgi, who had been at one time an assistant to Kepler, invented a system of logarithms very similar to that of Napier. In this case the successive terms of both the arithmetic and the geometric progressions increased. He started with zero and a common difference of 10, and with 100 000 000 and a common ratio of

$$1 + \frac{1}{10^4}$$

His calculations led to a table which started:

Arithmetic progression	*Geometric progression*
0	100 000 000
10	100 010 000
20	100 020 001
30	100 030 003
40	100 040 006
. . .	. . .

The final term of the arithmetic progression was corrected to give the number pair

230 270·022	1 000 000 000

Unlike the approach of Napier, Bürgi's was algebraic. Others, however, were working on similar lines. One of these others was Henry Briggs, then a professor at Gresham College in London. He travelled to Edinburgh to meet Napier and persuaded him that his tables would be more useful if the logarithm of 1 were to be 0 and the logarithm of 10, 1. This,of course, meant that two pairs of corresponding values were fixed, and a new form of computation was therefore needed for obtaining the tables. We now have

Arithmetic progression	*Geometric progression*
0	1
1	10
2	100
3	1000
4	10 000
...	...

but we need to calculate the interpolated values. To do this, Briggs used square roots.

To find the logarithm of 5, for example, we first note that this must lie between 0 and 1. We first take the logarithm of the square root of 10, which is just half the logarithm of 10, giving us

0·5 3·162 277

to seven places. The required logarithm of 5 must therefore lie between $\frac{1}{2}$ and 1. For the next stage we take the arithmetic mean of the logarithms and the geometric mean of the corresponding numbers, since the former belong to an arithmetic progression and the latter to a geometric one. This gives us

$$\tfrac{1}{2}(0{\cdot}5 + 1) \qquad \sqrt{3{\cdot}162277 \times 10}$$
$$= 0{\cdot}75 \qquad = 5{\cdot}623413$$

We now have a number greater than 5, so we take the arithmetic mean of $\frac{1}{2}$ and $\frac{3}{4}$ and the geometric mean of the corresponding numbers. This gives us

0·625 4·216 964

The number is now less than 5 and the procedure can be repeated until the logarithm of 5 is obtained to seven places. In fact, we obtain

0·698 970 0 5

Calculations of this kind have to be carried out for all the prime numbers, after which the logarithms of composite numbers can be found by adding the logarithms of their factors. Briggs himself published fourteen-place tables of logarithms of numbers from 1 to 20 000 and from 90 000 to 100 000. The gap was later filled by a Dutch publisher, Adriaen Vlacq.

These logarithms, which are to base ten, are known as 'common' or 'Briggsian' logarithms. They are the logarithms commonly taught in schools today, though the tables used are generally four-figure tables.

The principle by which these tables are compiled allow us to multiply or divide any two numbers by adding or subtracting their logarithms. Thus, to multiply two numbers a and b, we just use tables to look up their logarithms, add these, and convert the sum back from a logarithm to its corresponding number by reference again to the tables. In a similar manner, we can find powers and roots of numbers by multiplying or dividing their logarithms.

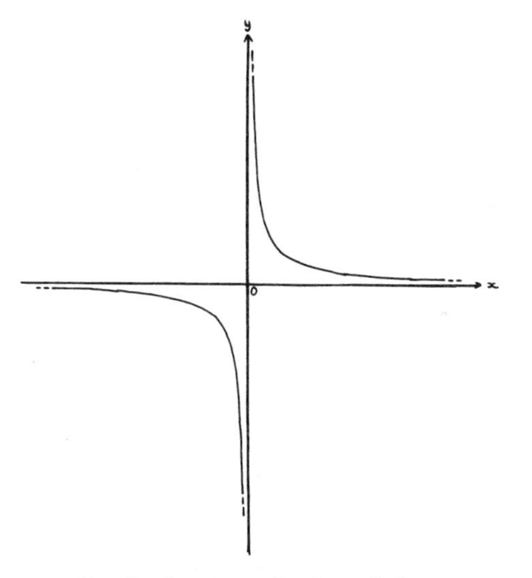

Graph of rectangular hyperbola

The logarithmic series

$$\log(1 + x) = x - \frac{x^2}{2} + \frac{x^3}{3} - \frac{x^4}{4} + \dots$$

where x must lie between -1 and $+1$, or be equal to 1, arose from a consideration of the area under the curve representing the relation

$$y = \frac{1}{x}$$

This curve, called the rectangular hyperbola (one of the conic sections, known since Greek times) was studied afresh in the seventeenth century by a number of eminent mathematicians, including Gregory of St. Vincent, Isaac Newton, and Nicolaus Mercator. It was from the logarithmic series, and others which converge more quickly, that tables of natural logarithms were computed. These natural logarithms, to base e, can be used in just the same way as logarithms to base ten. The basic principle is the same, namely that multiplication and division of numbers can be reduced to addition and subtraction of their logarithms. It is this principle which underlies the design of the slide rule.

Arithmetic Symbols

The modern signs +, plus, and −, minus, came into general use in Germany towards the end of the sixteenth century. Various suggestions have been put forward to account for their origin. Isolated examples of symbolic forms to denote basic arithmetic operations had appeared from time to time much earlier. There were, for example, the walking lege, Λ and Λ, Egyptian hieroglyphic forms for add and subtract, and the arrowhead ↑ for minus in the writings of Diophantus although he had no symbol for plus. In the fifteenth century the forms p̃ for plus and m̃ for minus were widely used in European mathematical manuscripts. The hypothesis has often been put forward that ~ was used in conjunction with the vertical stroke of p to give ⳁ and eventually +, and was used by itself for minus, eventually becoming just −. There are, however, alternative explanations.

It is possible that + derives from a shorthand form of the Latin word *et* meaning 'and'. According to this theory addition was first written in the Latin equivalent of, say, 'two and five equals seven'; the Latin *et* later became contracted to something like ⳁ, eventually becoming +. It seems unlikely that this hypothesis is correct because there are examples in various manuscripts of crosses being used, such as †, ⟶, ⟵, or even a Maltese cross ✠. However, one piece of evidence which can be claimed to support the theory is that in some manuscripts + is used for the general word 'and'. This can be explained away, of course, by suggesting that + as an abbreviation for *et* came after its adoption for addition simply because it was often spoken aloud as *et* just as we say today 'two *and* two makes four'. The evidence is therefore inconclusive.

An alternative popular theory for the derivation of the subtraction symbol − suggests that the letter m̃ standing for minus came to be written somewhat carelessly as little more than a squiggle ~ which in turn became the horizontal line —. Other theories seek its origin in a sign used by merchants to denote the difference between the gross and nett weights of packaged goods, or in a corruption of Diophantus' ↑ through an intermediate form ⊤ from which the vertical stroke eventually disappeared. However, contemporary with the early use of —, we find examples of our division symbol ÷ being used to denote subtraction. Since a horizontal bar was used in some manuscripts to denote proportion or even equality, we can reasonably infer that the form ÷ was adopted temporarily for subtraction simply to avoid confusion with these other uses of —. As with the addition sign, the evidence for the true origin of − for minus is inconclusive.

The first appearance of our × for multiplication is usually said to be in a seventeenth-century work by Willian Oughtred, though there are also claims that it dates from the previous century. There had been symbols for 'times' as far back as the Babylonian period, and in early documents from India we occasionally find the use of • familiar to us today. It seems possible that the St. Andrew's cross came into general use as a result of writing cross multiplications. Here, four numbers are involved and not just two. We have, for example,

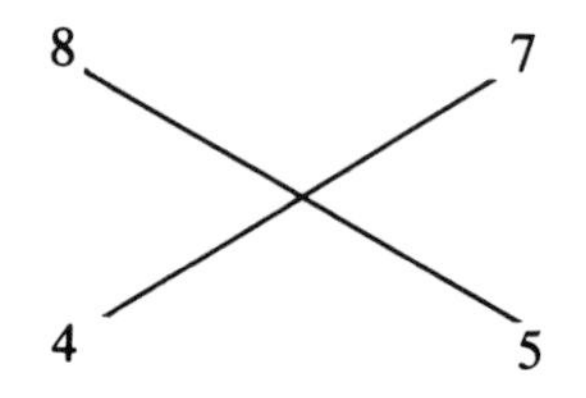

where 8 and 5 are first multiplied together and then 4 and 7, the products being either added or subtracted according to the particular calculation involved. The St. Andrew's cross was, however, used for other purposes, usually associated in some way either with proportion or with the rule of false (see Chapter 6).

Occasionally, we find the symbol ∩ denoting multiplication. This is clearly derived from the letter m and was introduced much at the same time as it became popular to use x to stand for the unknown quantity in algebraic equations. The use of ∩ avoids all possible confusion, especially when around the same time simple juxtaposition of letters had become quite commonplace, ax meaning the number a times the unknown x. An expression such as axy could be misinterpreted, and even more so an expression such as axx when xx was still being used in place of our more familiar x^2 to denote the square of the unknown. We must also remember that it was by no means the custom to be scrupulous about printing algebraic expressions in italics. The use of ∩ was, however, shortlived and the general practice became the use of × in arithmetic and juxtaposition of letters in algebra.

We can be rather more certain about the origin of the symbol ÷ for division. The Greeks had denoted fractions by writing the denominator above the numerator, as for example $\frac{\varepsilon}{\gamma}$ meaning three-fifths, and later by writing the numerator over the denominator, three-fifths then becoming $\frac{\gamma}{\varepsilon}$. The Hindus wrote fractions much as we write them now though, like the Greeks, without the horizontal bar which seems to have been a later Arab innovation. Writing fractions such as $\frac{3}{4}$ to denote three divided by four in the course of a written sentence involves either using up additional vertical spacing or writing very small numerals. Clearly, it is desirable if possible to have the two numbers horizontally aligned with the rest of the writing, and we saw how this was accomplished by the use of the solidus or stroke. However, it is only a small step from, say $\frac{9}{3}$ to 9 ÷ 3, the dots indicating the positions originally occupied by the numerals when written in fractional form, and also serving to provide a clear distinction between division and subtraction. Sometimes we find more than one dot written above and below the bar line. It seems reasonable for us to assume that this practice arose from division of numbers having more than one digit. Thus, if $\frac{9}{3}$ became 9 ÷ 3, $\frac{17}{12}$ could by analogy become 17 ∺ 12 and $\frac{17}{5}$ could become 17 ∹ 5, and so on. Forms of division symbol with unequal numbers of dots above and below the line are to be found in a few mathematical works. However, we do find different symbols for division as with the eighteenth-century Ↄ in France, clearly just the initial letter of the word 'division' with the capital D reversed.

The ideas underlying ratio and proportion are, of course, closely allied to the operation of division. We should therefore expect to find similar symbols used in each case, and indeed this is just what often happened. The forms : and :: have been widely used to denote ratio especially in Europe and America. Thus, a typical example is

$$2:3::4:6$$

Since this can be interpreted also to mean that 2 divided by 3 is equal to 4 divided by 6, there is an apparently straightforward link between the use of the colon in writing down a ratio and the use of the colon combined with the horizontal bar to denote division. In fact, there is no historical justification for this link, attractive though the idea may be. We do, however, find isolated examples of the use of the colon alone for division. In the sixteenth and seventeenth centuries especially, there was considerable variation in the notation used for ratio, as with many other arithmetic symbols. We find single, double and triple dots, single and double strokes, both slanting and vertical, and even the use of the Greek letter π. It would seem that it was the need to make a clear distinction between ratio and division, just as between division and fraction, which eventually led to special and distinct symbols being used for each.

In addition to the symbols for the basic operations there are many other symbols in general use such as our familiar symbol = for equality, which appeared in the writings of Roberte Recorde (sixteenth century) who claimed that 'no two things can be more equal than a pair of parallel lines'. Previous to Recorde's time it had been customary in the West either to write out the appropriate word in full or to use some form of abbreviation such as *aeq.* for *aequantur*. We find a temporary seventeenth-century rival of = in the ∝ of Descartes, claimed as a further abbreviation of *ae*, but possibly the astronomical symbol ♉, taurus, turned on its side. Earlier symbols used include an Egyptian hieratic abbreviation usually translated as 'gives', the letters ι^{σ} of Diophantus, ∴ in some Arab works, and in western Europe a dash or, occasionally, 2|2.

Special symbols for roots were used from ancient times such as the symbol ⌐ in Egyptian texts. In the West the Latin word *radix* was widely used in translations made from Arabic works and, by the sixteenth century, the letter R, the first letter of *radix*, was very common. This was eventually displaced by our √ which was invented in Germany and then gained ground steadily throughout the seventeenth century. Its exact origin, however, is uncertain. For a long time it was thought to be a kind of shorthand form of R or even a form of reversed ℓ, *l* being the first letter of the Latin word *latus*. Another possibility is that it evolved from a practice of indicating a square root by a dot. This, it is claimed, came to be modified in order to avoid possible confusion with other meanings of a dot, for example multiplication, the modification being the addition of a tail. Thus, the square root of a number was indicated by writing ∫ in front of it, and this eventually became √. Once again, we find more than one theory of origin, and those supporting each theory can find data, though inconclusive data, to support their particular arguments.

As with roots, various special symbols have been used from ancient times to denote powers. Powers are found written most frequently in algebraic expressions. Diophantus developed a notation in which the square of an uknown number was indicated by Δ^{Y}, its cube by K^{Y}, and powers up to the sixth power by combinations of Δ and K, the fifth power, for example, being denoted by ΔK^{Y}. This notation is additive, and in this sense it compares with our writing of the fifth power of the unknown as x^{2+3}. Diophantus' notation is just the use of first letters of *δυναμις* (power) and *κυbος* (cube), either in capitals, as in our examples, or in small letters. In Latin translations of Greek and Arab works and in original works written in Latin, we often find Q (from *quadratus*) and C (from *cubus*).

From the fifteenth century onwards we find conventions which are similar to our modern system of superscripts by which we write the nth power of x as x^n. In some instances we find the superscripts written in Roman numerals so that we get expressions such as 3^{iv} for what we would write as $3x^4$. It was quite common to omit the symbol for the unknown quantity, thus, even after the Roman superscripts had been abandoned, we find the same quantity written as 3^4. The important step of using letters to denote both constants and unknowns was taken by Francis Viète in the sixteenth century. He used consonants for constants and vowels for unknown numbers, though still retaining expressions such as *E quadr*. for the square of a number. The final stage of using small letters from the beginning of the alphabet for constants and from the end of the alphabet for variables is due to the great René Descartes who, though he sometimes used xx for the square of the unknown, in general used notation very much as we do today. Descartes' *Géométrie* has thus often been described as the first modern textbook. Apart from the appearance of ∝ for equality, we can read all the notation employed in it almost without noticing that it is not a twentieth-century work.

Chapter Five
Computing with Numbers

I cannot do't without counters.
(Shakespeare)

It is the Age of Machinery, in every outward
and inward sense of that word.
(Carlyle)

In Chapter 4 we discussed ways of carrying out the basic operations of arithmetic, confining ourselves to methods which did not involve any kind of special apparatus or machine. In this chapter we consider physical aids to calculation: the abacus, the slide rule, and the modern computer, and we use 'computing' in its title, having used 'calculating' in the title of Chapter 4. This is a somewhat arbitrary choice of words. In the past, the words 'compute' and 'calculate' have been largely synonymous. Today, however, they have slightly different shades of meaning: if we speak of a computer, we shall certainly be understood to be referring to some kind of machine. Of course, if we speak of a calculator, much the same will be true also, though the latter will probably be taken to denote one of the pocket calculators now readily available on the market. Such pocket calculators carry with them no overtones of menace – they are seen as a convenient way of avoiding the drudgery of simple arithmetic! Computers, on the other hand, are regarded with suspicion by the layman, if for no better reason than that the computerized society of the not very distant future is seen to be one which is unacceptably impersonal and presents a serious threat to jobs, privacy, and individual initiative. We shall not be considering the social implications of modern developments in computing here. We shall be concerned only with the role of computers in performing arithmetic operations.

Early Counting Boards

The most common methods of computation in ancient times involved the use of a counting board or abacus. Unfortunately, the precise details of many of these have been lost to us. We do have a few surviving ancient counting boards, but we have to be content with making intelligent guesses when it comes to reconstructing the calculations performed on them. Indirect evidence is provided by surviving written calculations, by treatises from later times, and by occasional references in contemporary literature, for example, in Herodotus and Polybius.

The origin of the counting board lies in the boards covered with dust or sand in which marks could be scratched using the tallying principle. These came into being because of the inconvenience of dealing with larger numbers on the fingers. Soon, pebbles were used for making simple calculations and the boards came to be marked out with permanent markings indicating rows or columns. Later, grooves were cut in wood or stone in which small spheres could be moved up and down, and later still the spheres were pierced or replaced by beads. This meant that they were now permanently attached to some kind of frame, and were moved up and down on wooden or metal rods.

Essentially the counting board, throughout its history, has been used mainly for purposes of trade and keeping of accounts. Written calculation, especially with some of the numeral systems of ancient times, was an advanced and specialized art often confined to the priests or to other highly-educated classes. The counting board by contrast was the preserve of the traders and merchants, the tax-collectors, and the many other local officials. Inca officials were a typical example of this. They had no written system of numerals. All their calculations were performed on counting boards. Maize kernels rather than pebbles were placed on these to represent numbers and the results of calculations were then recorded on their quipus.

We know what the Greek and Roman abaci looked like. There is a surviving Greek white marble counting board, roughly $2\frac{1}{2}$ by 5 feet, discovered during excavations in the nineteenth century on the island of Salamis. We also have two surviving Roman hand abaci. Pictures of abacus calculation can be seen on a vase from the fourth century before Christ recovered in southern Italy, on an Etruscan cameo of uncertain date, and on a first-century gravestone. The principle of the ancient form of the counting board can be deduced from this evidence.

We can think of the counting board as a set of parallel columns representing the various powers of ten together with some means of making entries in these. A typical ancient decimal counting board thus consists of columns headed in numerals corresponding to our 1, 10, 100, 1000 and up to nine pebbles or counters available for each column. The Romans called such pebbles *calculi*, from which we derive our word 'calculate'. Similarly, the Greek words for 'stone' and 'calculation' have a common root. Our 3821 would be represented by:

o	o	o	o
o	o	o	
o	o		
	o		
	o		
	o		
	o		
	o		

where the column headings would be (from right to left) I, X, C, CIↃ or I, Δ, H, X for the Roman and Greek abaci respectively.

The addition of another number, say our 1495, is carried out simply by adding further counters. We start by adding 5 more in the units column, and we follow this by adding 9 more in the 10s column. This gives us 11 in this column, so we remove 10 of these and replace them by 1 in the 100s column. We next add 4 counters to the 100s column making 13 in all, of which 10 are removed and replaced by an additional counter in the 1000s column. Finally, we add a further counter to the 1000s column, the final sum, 5316 appearing as

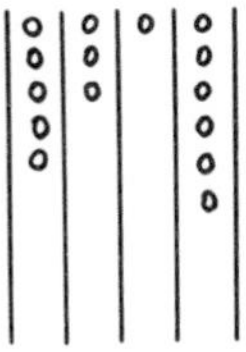

Subtraction is carried out in much the same way. When in any particular column we need to subtract a larger number from a smaller one, we just remove a counter of the next higher denomination giving us 10 further counters in the column in which we are working.

The Greek Attic and the Roman numeral systems both included collective symbols for groupings of 5. These intermediate units were introduced into the counting boards. Originally each column was divided vertically, so that the headings on a Greek board would now be

𐅈 Χ 𐅇 Η 𐅆 Δ Γ Ι

The Salamis tablet shows us that special columns were also included for fractions. However, the Romans introduced a horizontal line crossing the decimal columns, counters above the line counting as 5 corresponding units. Our previous total of 5316 would thus appear as

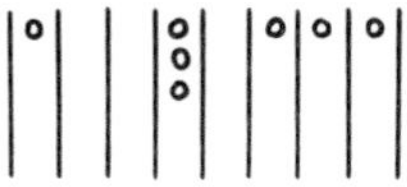

on a Greek board, and as

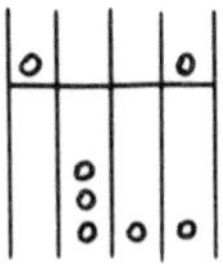

on a Roman board. The advantage here is that fewer counters are required and, for multiplication and division, only products up to 5 × 5 need to be learned or read from tables.

With the large boards it was possible to enter more than one number at a time. This was useful, particularly for the purposes of addition. However, the Roman hand abacus did not allow for this. The spheres required to denote numbers remained all the time in their grooves and were moved against the dividing bar as required. The same number which we represented above

would thus appear as

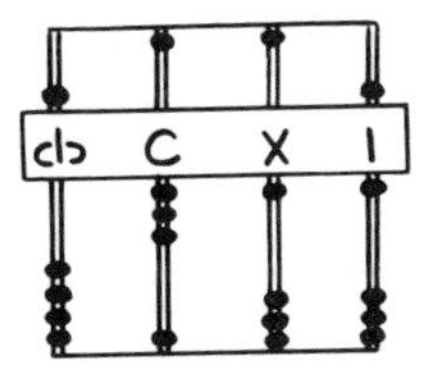

In carrying out multiplication, it was customary to begin with the highest powers. There was one difficulty in this – multiplication tables up to 5×5 could not on their own provide the necessary information as to which column should contain the resulting product. For example, multiplying 300 by 500 gives a result of 150 000. Tables, however, could only supply the partial answer, 15. In order to know where the appropriate entry should be made, the calculator needed to know an additional special rule. This stated that the units figure of the product should be entered in the pth column from the right, where

$$p = a + b - 1$$

a and b being the furthest columns from the right needed in the representation of the numbers being multiplied. Thus, for 300 and 500 the 5 of their product would be entered in the fifth column, that is, the column headed ↂ (10 000). The 1 of the product would thus be entered in the next column, that headed ↈ (100 000).

It is clear that the Roman abaci must have survived in one form or another after the collapse of the Roman Empire. What seems most probable is that they continued to be used in Christian monastic communities. However, we have no further record of the details of calculations on the abacus until the latter part of the tenth century, by which time a highly significant change had taken place in the nature of the counters.

Counting Boards in the Middle Ages

The new form of the counters, introduced in the tenth century, was due to Gerbert, later to be Pope Sylvester II. He was born of poor parents in the Auvergne in the year 940. He was given a monastic education and later travelled to Spain, where he became acquainted with the Hindu–Arabic numerals. He has left a description of how arithmetic calculations were carried out on his abacus, but for a description of the abacus itself we have to rely on the writings of his pupils.

The board of Gerbert's abacus was similar to the ancient decimal Roman counting boards. There were columns headed with the various powers of ten in Roman numerals. Arches were drawn over the top of each column. These were called *arci Pythagorei* because at the time it was thought, incorrectly, that Pythagoras was the inventor of the counting board. Instead of using identical counters on the tallying principle, however, Gerbert introduced counters on which Hindu–Arabic numerals had been written. These counters were known as *apices*. 5316 would now appear as

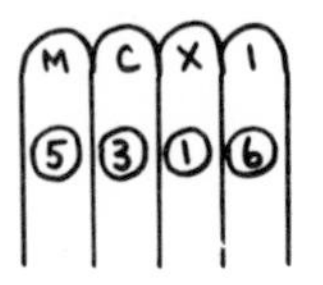

(using modern numerals).

As far as computations on the abacus is concerned, Gerbert's innovation was a near disaster. Before any calculation could be started the apices had to be sorted, and at each and every stage of the calculation apices had to be replaced by others of different denomination. This was not too much of a problem with addition and subtraction, although the visual assistance of the tallying principle was entirely lost. For multiplication and, especially, division the problems were serious. To see this, we shall follow through the stages of two calculations, multiplication of 163 by 82, and division of the product, 13 366 by 82.

The multiplication would first be set up as follows:

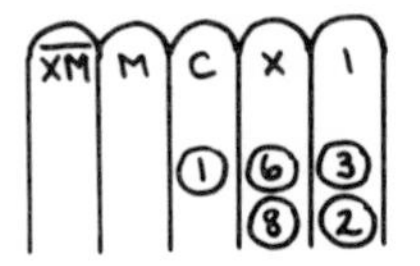

Multiplying 1 by 8 yields 8, and the rule for locating the appropriate column yields $3 + 2 - 1 = 4$, so an 8-counter is placed in the column headed M. 6×8 yields 48, and the same rule requires us to place an 8-counter in the column headed C. When we come to enter a 4-counter in the M–column, however, we find that this is already occupied. We therefore have to add 8 and 4, giving 12, and enter a 2-counter in the M–column and a 1-counter in the $\overline{XM}$–column, $\overline{XM}$ being the usual heading at this period for the column earlier headed C|Ɔ. The abacus now reads

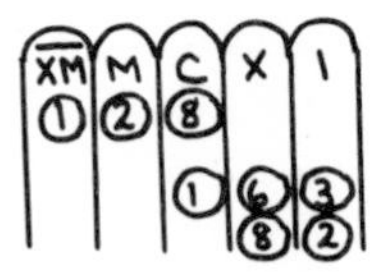

3×8 now yields 24. A 4-counter has to be placed in the X–column, then the 8-counter in the C-column has to be removed and the 2-counter in the M–column replaced by a 3-counter. This completes the multiplication by 8, and the 8-counter of the multiplier can be discarded. We are now in a position to commence multiplying by 2. This follows a similar process, counters having to be removed and replaced at each stage of the remaining part of the calculation except the final one, where 2 multiplies 3 to give 6. The 2-counter of the multiplier can now be removed along with the counters representing 163, and

the final product appears as

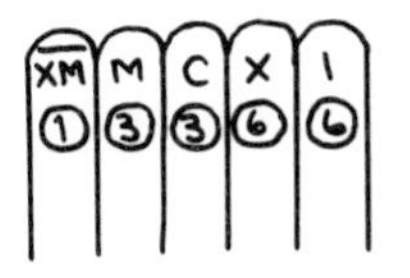

We can see just how tedious the perpetual exchanges of counters were.

With division, the situation was particularly complicated. Gerbert describes two methods, known respectively as 'golden' and 'iron' division. Golden division is in many ways similar to our method of long division, and there should be little difficulty in following through the calculation. We have struck out the counters which are removed at each stage, and the counters replacing them appear immediately below.

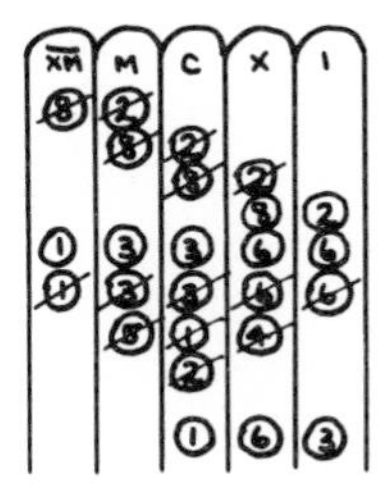

We should note that the counters representing the divisor are moved one place to the right at each stage, that those representing the number being divided remain unchanged throughout the calculation, that remainders are indicated successively immediately below these counters, and that the quotient counters are placed at the bottom. Clearly, a certain amount of mental arithmetic is required in obtaining the remainders. The rule of places used in multiplication was adapted to indicate where the first counter of the quotient should be placed. In our example, the numbers divide exactly. A remainder, if any, would be represented by remainder counters not finally removed by the last division.

Iron division was so called because it had the reputation of being 'harder than iron' to carry out. In our example, the process begins by increasing the value of the divisor to 90, the nearest multiple of 10 greater than 82. We begin, therefore, with the abacus appearing as

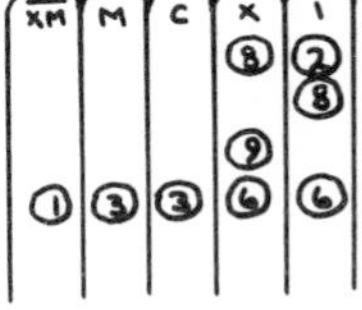

The first figure of the quotient is obtained by dividing 13 by 9, and using the place rule to indicate that a 1-counter should be placed at the bottom of the board in the C–column. The 1- and 3-counters in the $\overline{\text{XM}}$– and M–columns

are now discarded, and replaced by a 4-counter in the M–column representing the remainder. We now have

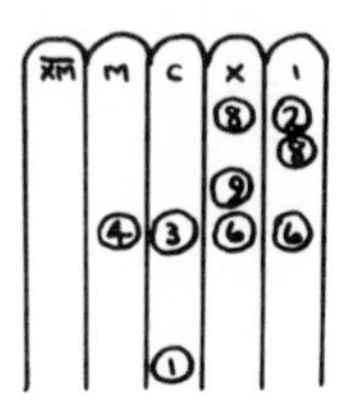

However, we have divided by 9 and not 8, there is therefore a correction to be made. This is done by multiplying the increase of the divisor, 8, by the 1 of the quotient obtained so far, and increasing 4366 by 800. This line is now changed so as to read

We now divide 51 by 9, and put a 5-counter in the X–column at the bottom. The 5- and 1-counters in the M– and C–columns are now removed, and a 6-counter, for the remainder, placed in the C–column. 666 has, however to be adjusted by the product of the divisor increase and the new quotient figure. This means adding 400, so that we now have

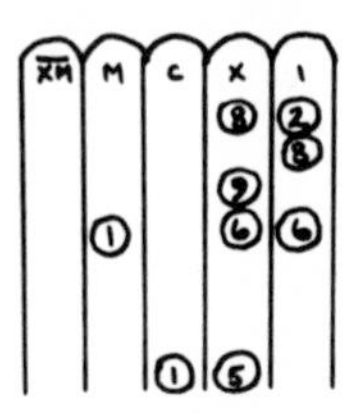

Continuing the same pattern of calculation gives a quotient of 11. We therefore alter the quotient at the bottom to 161, calculate the remainder as 7 in the X–column, and add in the adjustment of 88. We have therefore

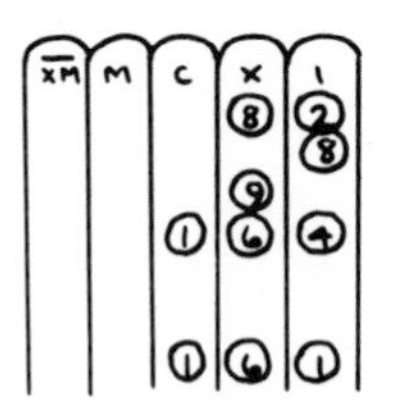

Dividing again by 9 gives a quotient 1 and remainder 7. We increase the overall quotient by 1, enter the new remainder, and adjust by adding 16 to 74 to give

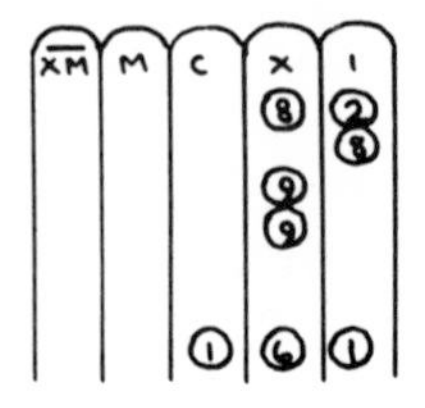

Finally, dividing again by 9 makes us adjust the quotient to 162 and the calculation is complete. A final remainder, if any, would appear at the end in place of the original number being divided. It is small wonder that this method was given its uncomplimentary title!

Gerbert's form of the abacus is described in several extant works in which there is not always agreement as to the forms of the numerals on the apices. Confusion about the Hindu–Arabic numerals arose at this time because the apices could easily be rotated when placed on the board, so that 2, for example, could just as easily appear as ᓄ, ᘔ, or ᓇ.

It is doubtful if Gerbert's abacus was ever used extensively outside the monasteries. It failed as a means of introducing the new numerals because Gerbert had been mistaken in thinking that these were appropriate for abacus calculation. The idea of abacus calculation, however, passed to the merchants and traders, but they reverted to the tallying principle, rotated the board through ninety degrees and restoring the groupings of 5.

Reckoning on the Lines

Reckoning or 'casting' on the lines, as it was often called, developed from the twelfth century onwards and became the normal method of computation for trade and money exchange generally. There was considerable expansion in trade in this century, largely as a result of the Crusaders and the activities of the Hanseatic League. Gerbert's abacus was unsuitable for widespread use because the new numerals were not widely known and, in any case, their introduction was often strongly resisted. The principle of decimal columns was replaced by that of horizontal decimal lines. Plain counters were now placed actually on the lines, which might be drawn on paper, embroidered on cloth, or engraved on wooden tables. Occasionally a plain cloth would be used and counters or other objects placed one below the other at one side to indicate where lines would have been had they actually been marked out. Up to 4 counters were placed on any one line, the space between the lines being used for counters representing 5 of those on the line immediately below. The number, 5812 would therefore appear as

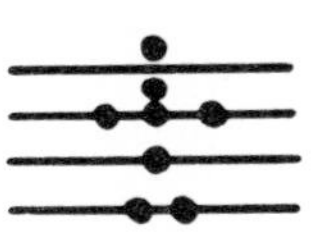

Sometimes the lines would be designated with symbols representing a specific coinage. Often they would be completely unmarked as in the example shown above.

There was one advantage of this new orientation – it was now easy to represent several numbers side by side. However, the parallel with the new place-value notation, gradually being introduced, was lost, and those who practised reckoning on the lines became fierce opponents of the Hindu–Arabic numerals and the new arithmetic associated with them.

There are many textbooks which describe exactly how calculations were performed on the lines. The operations discussed include numeration, addition, subtraction, duplation, mediation, multiplication, division, elevation and resolution. Numeration simply meant placing counters on or between the lines. Duplation and mediation were the operations of doubling and halving. Elevation was the process of grouping counters together so as to replace them with counters higher up the board. Resolution was the reverse of elevation, that is, it involved the decomposing of a counter into either 2 or 5 counters, as appropriate, of the next level below. A typical calculation would involve several of these operations.

Suppose that we wish to subtract 278 from 364. We first set up the two numbers side by side:

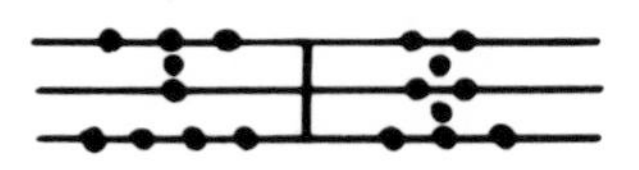

This completes the initial operation of numeration. We now carry out sufficient resolution to make immediate subtraction possible at each level:

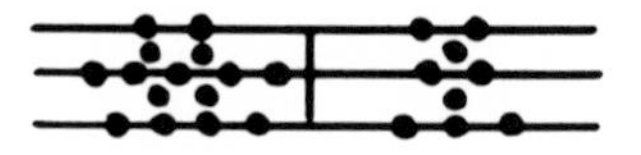

Then we perform the operation of subtraction, that is, at each level we remove as many counters from the left as there are counters on the right. We are left with

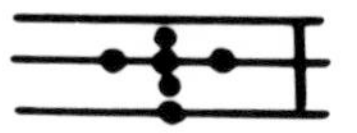

representing 86, the correct result.

The operations of duplation and mediation can quickly be carried out according to simple rules. For duplation of counters on a line, we just add as many as are already there, carrying out elevation where appropriate. Duplation of a counter in a space involves moving it up to the line immediately above. Mediation of counters on lines involves removing half of them if there is an even number and moving one counter down into the space below because of the one left over when halving an odd number. Halving the number represented by a counter in a space demands resolution since half of 5 involves the fraction $\frac{1}{2}$. However, as an alternative method, 2 counters can be placed on the line immediately below and 1 counter in the space below that.

In England, a special kind of counting board came into common use in government offices. It was used in addition to the usual form of reckoning on the lines. The board was divided horizontally and vertically, the vertical divisions representing (from right to left) pence, shillings, pounds, and twenty-pounds. Up to 19 counters could therefore be required for any of the columns except the one on the right, for which up to 11 counters could be required. A clever system was adopted whereby a counter placed above the left-hand end of a group of counters in any particular cell, other than the pence cell, meant 10, and one placed at the right-hand end meant 5. In a pence cell, a counter placed above the right hand end of a group meant 6. The amounts specified in the two rows are therefore equivalent to £197 13s 6d and £74 18s 11d.

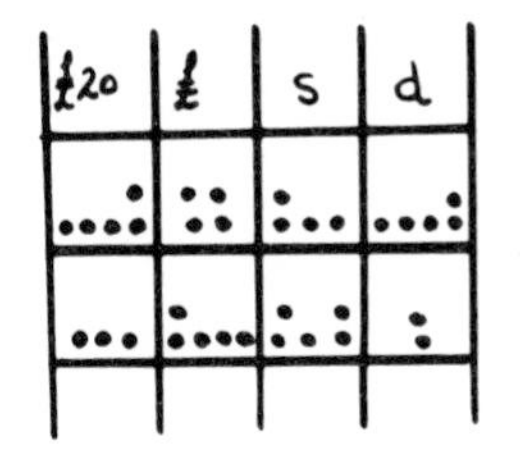

Patterns of counters were copied down directly into documents, though this seems to have happened only rarely. The main reason for this is that money boards using counters in this particular way were introduced in the sixteenth century, by which time written calculations using the Hindu–Arabic numerals were beginning to supersede the various forms of counting board throughout western Europe. By the close of the seventeenth century the counting board had almost completely disappeared, so much so that (as we saw in Chapter 3) a Russian abacus brought to France after the Napoleonic campaigns was regarded as a remarkable curiosity.

The abacus and reckoning on the lines did, however, leave behind a permanent heritage of words. We have already seen that our word 'calculate' is one of these. Our word 'bureau' formerly denoted a table whose top was engraved as a counting board; today it denotes a writing table, written calculations having replaced reckoning on the lines. The true origin of 'bureau', however, comes from a Latin word for a skein of wool, which later came to be adapted to mean a woollen cloth, and still later a reckoning cloth. The words 'exchequer', 'cheque', 'chess' and 'check' all derive from the chequered board used for computation. Similarly, the words 'bank', 'banker', and 'bankrupt' derive from the old German *Rechenbank* (counting board), bankrupt meaning literally a broken board. Similar words have passed into general usage in many of the European languages.

The Modern Abacus

Although the Roman hand abacus never became widely popular in the West, it nevertheless penetrated eastwards, and modern forms of it are still extensively used in the Far East. It was adopted in China, probably in the twelfth century, and passed to Japan in a modified form some four centuries later. It then returned westwards and linked up with an old tradition in the Slavic countries where it can still be found. Displaced persons brought the Slavic form back into western Europe, where it can still occasionally be seen in practical use. We are probably most familiar with it from the classroom, where it has frequently been used for teaching elementary arithmetic to young children. Special forms of abacus have in fact been developed for teaching purposes, of which the abacounter is an example.

The Roman hand abacus had grooves in which small spheres could be moved up and down. However, the Chinese form always seems to have consisted of beads on bamboo rods. It is possible that this change was suggested by the knot-numbers, which were widely used in the Far East. The *suan-pan*, as the Chinese abacus is called, usually has 2 beads, representing fives, on the upper part of each wire and 5 beads on the lower part. Beads are moved towards the dividing bar, which is said to separate heaven (the fives region) from hell (the units region). There are usually sufficient rods to enable at least two 4-digit numbers to be entered side by side. Thus, 53 and 2178 appear on the *suan-pan* as

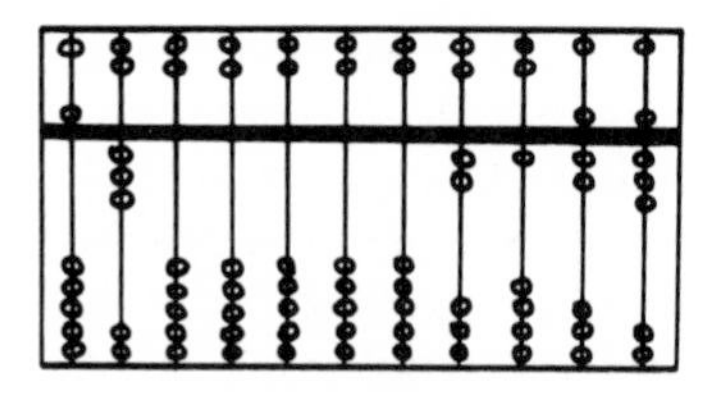

Although strictly only 4 beads are needed for units and 1 bead for fives, the calculation methods employed do make use of the additional beads available. Numbers up to 15, for example, can be represented on just one wire by making use of all the beads available on it. Before the adoption of the abacus, the Chinese used small sticks, called *sangi*, on a counting board. It was this practice which was associated with the stick numerals discussed in Chapter 3.

The Japanese *soroban* usually has even more rods than the *suan-pan*. Generally, there will be 5 beads for units and only 1 for fives. A little of the versatility of the *suan-pan* is therefore lost. However, the Japanese have developed an incredible dexterity in using the *soroban*. When the Americans occupied Japan in 1945, they were inclined to regard the *soroban* as evidence of its backwardness. To demonstrate the superiority of the American electrically-powered calculator, a competition was organized in Tokyo between a Japanese abacus expert and an American army clerk experienced in the use of the automatic calculator. In all the operations attempted, except for multiplication, the Japanese abacus proved superior both in speed and accuracy.

The *suan-pan* and the *soroban* are still to be seen today. They are used in both elementary and secondary schools, where the children make surprising progress in arithmetic. They are used in shops, in banks, in accountants' offices, and in government finance departments, though modern computers are also extensively used for the processing and storage of large-scale information. Essentially, the abacus has remained the hand-calculator for the individual, and it remains to be seen for how much longer it will retain its popularity against the rapidly increasing challenge of the pocket electronic calculator.

The Slavic form of the abacus is normally used with the wires in the horizontal rather than the vertical position. There are ten beads on each wire, the middle two beads being coloured differently from the rest. This proves an even more effective way of allowing for variations in representing numbers during calculations. In some cases, particularly where the abacus is used primarily for accounting purposes, the bottom and fourth-from-bottom wires each have only four beads, though the middle two still retain a distinctive colour. This allows for calculation in quarters corresponding to the denominations of the currency. It was this type of Russian abacus which found its way into France in the nineteenth century and was eventually adapted in a purely decimal form for use in the elementary schools of western Europe.

The Slide Rule

Three years after he had published his first discussion of logarithms, John Napier described another invention which was eventually to have a profound influence on the way in which calculations were carried out, especially in engineering and the sciences. This invention, known as 'Napier's rods' (or, sometimes, 'Napier's bones') was the forerunner of our modern slide rule.

The principle of Napier's rods is based on the method of lattice multiplication. Each rod is headed by a numeral n and below it are its various multiples up to $9 \times n$ entered in diagonally divided cells. Suppose now that we want to multiply 53 by 7. We first place the rods headed by 5 and 3 side by side.

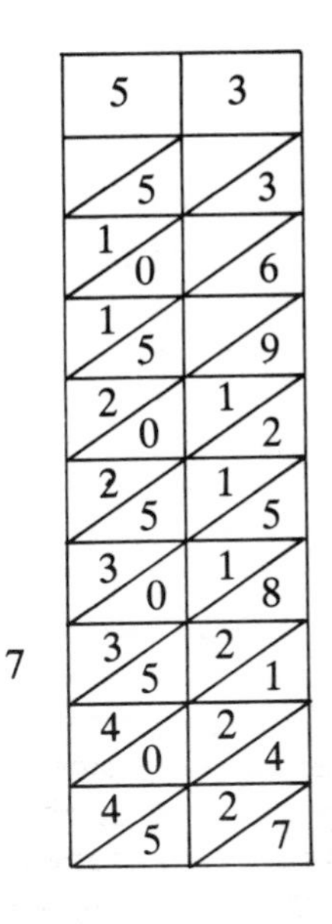

We read off the required product along the seventh row as 371, the 70 being obtained by adding the 5 and 2 as in lattice multiplication. If we wish to multiply by a number greater than 10, we multiply by its digits separately and add the results obtained. Thus, in multiplying 53 by 74, we read off 371 and

212 and carry out the addition:

$$\begin{array}{r} 3710 \\ 212 \\ \hline 3922 \end{array}$$

Napier's rods became very popular in the seventeenth century, and various kinds were manufactured. Usually, they had a square cross-section allowing numbers to be written on all four sides. The rods were then headed on each side, so that their contents could be known at a glance. Thus, a rod on whose sides were the multiples of 2, 4, 6 and 8, would be headed, for example, on one of its sides

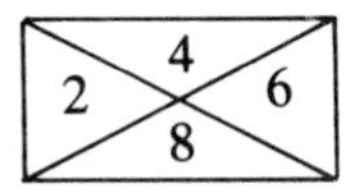

In this case, the multiples of 8 would be uppermost, those of 2 and 6 to the left and right, and those of 4 on the bottom.

It was just as easy to carry out division. The multiples of any divisor could be seen immediately. Thus, if we are dividing 3922 by 53, we set up the rods as we did when 53 was our multiplier, and first note that the seventh row reads

3 / 5	2 / 1

We now perform the subtraction

$$\begin{array}{r} 3922 \\ 3710 \\ \hline 212 \end{array}$$

and then note that the fourth row reads

2 / 0	1 / 2

The quotient is thus 74.

Various developments in the design of the rods were quick to take place. The original rods were designed to be placed side by side. However, cylindrical rods appeared with multiples of the numbers 0 to 9 on each rod. They were mounted in a frame and rotated independently as required. Unfortunately, it was more difficult to read along the rows as the operative columns were no longer touching each other. Napier himself extended the use of his rods to other mathematical operations, though, of course, the numbers on the other various columns had to be altered accordingly.

Once the principle of calculation by placing rods side by side became well-known, it was not long before the idea of rods sliding against each other was adopted for computation. Some five years after Napier had published the work describing his rods, the slide rule was invented by Willian Oughtred, though he did not publish the details for another ten years.

The simplest form of slide rule is one on which we can perform addition and subtraction. Here, we need two rules marked arithmetically. We can then read off directly the addition of, say, 28 and 43. What we are doing, of course, is just adding lengths proportional to the two numbers.

Subtraction is just as easy: we move the rules so that the two numbers are aligned and read off the required answer against the 1 on the rule carrying the larger number.

To perform multiplication and division simply by adding and subtracting lengths the rules have to be marked, not arithmetically, but so that distances are proportional to the logarithms of numbers. This idea had been pioneered by Edmund Gunter, who, in 1620, had constructed such a logarithmic scale and performed multiplications and divisions by adding and subtracting lengths using a pair of dividers. Oughtred developed this idea by carrying out the same operations using two logarithmic rules sliding along each other. Thus, the slide rule evolved out of applications of two of Napier's inventions, logarithms and calculating rods.

To carry out multiplication on the slide rule we follow exactly the process we used for addition using the arithmetically marked rules. Thus, 2×3 is read off immediately as 6:

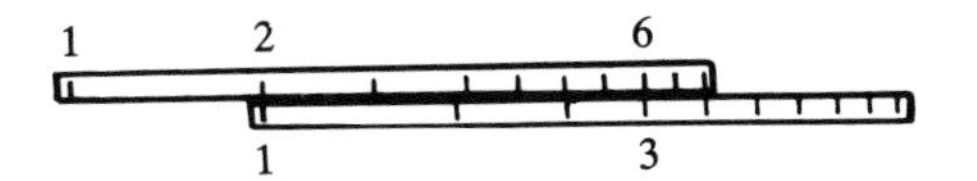

For division, we simply subtract lengths. If we are in difficulty because the number we wish to read off is outside the compass of the rule, we merely change from one as our base to ten, or any power of ten. This is made possible because the lengths are proportional to logarithms to base ten. All we are doing is putting into practical effect the two expressions:

$$\log(a \times b) = \log a + \log b$$

$$\log(a \div b) = \log a - \log b$$

Large numbers and decimal fractions are no problem. We use the numbers marked on the rules irrespective of any power of ten by which they may be multiplied. So, if we are multiplying 30 by 20, we just multiply 3 by 2 on the rules and read off the result obtained as 600 instead of 6. If we are multiplying 0·3 by 0·02, we read off the result as 0·006.

If we construct one of the scales so that its overall length is exactly half that of the other, we can immediately read off squares and square roots. This is just an application of:

$$\log a^2 = 2 \log a$$

If we align unity on these two rules, numbers and their squares will be aligned throughout. The usual modern slide rule has at least two pairs of rules, one pair constructed on just this principle. There is a central slide or tongue and a sliding cursor on which a hair line enabling readings to be taken with accuracy. The tongue allows for the two pairs of sliding scales, and was first incorporated around 30 years or so after the invention of the slide rule.

Various other scales were introduced in the first half of the nineteenth century, the final form of the modern slide rule becoming pretty well standardized by around 1850. Not all the slide rules were as we have described them, however. One of the problems was that restricting the rule to a manageable length meant that the two pairs of logarithmic scales were usually constructed reading from 1 to 10 and from 1 to 100. A scale reading up to higher powers of ten could not have been read throughout with sufficient accuracy. This difficulty was overcome by the use of cylindrical rules with spiral scales, usually called 'circular' slide rules. These were first described by Oughtred some nine years after his invention of the straight rule.

Calculating Machines

The next obvious stage in the development of aids to calculation was that they should become mechanized. The first machine for performing the four basic arithmetical operations was designed and constructed by Blaise Pascal in 1642 when he was only nineteen. We have a full description of this machine and a number of authentic models have been preserved. Essentially it was an adding and subtracting device; multiplication and division had to be carried out by successive addition and subtraction.

The machine consisted of a number of geared wheels set inside a box in whose top there were a row of windows for reading off the result of any calculation. Numbers were set into the machine on a row of wheels below the windows. The crucial part of the invention was the automatic carrying mechanism. A ratchet was introduced between adjoining wheels so that whenever a wheel passed from reading 9 to reading 0, the wheel next to it was moved on one digit. To enable this to be effective, the wheels had to rotate in one direction only; hence, subtraction was carried out by the addition of complements. Thus to subtract 8 we simply add 2 and deduct 10 from the result.

The mechanization of multiplication and division was achieved by Leibniz about 30 years after Pascal's invention of his machine. We have a full description of this later machine and some authentic models also. Leibniz's machine had two parts. One part, that for carrying out addition and subtraction, was similar to Pascal's; the other part was new. There were three kinds of wheel: wheels for addition, wheels for multipliers, and wheels for numbers to be multiplied. The crucial feature here was the introduction of gear wheels on which the number of operative teeth could be varied. These stepped wheels continued to be used in mechanical calculators until only recently, when the manufacture of such calculators ceased because of the competition from electronic machines.

Small and not so small desk calculators began to be manufactured on a commercial scale early in the nineteenth century, and by the end of the century they were commonplace in engineering and accountancy. Various refinements were introduced, including additional mechanism for printing the results of calculations. However, these machines were not in any true sense automatic. The various operations had to be effected by hand, often in several stages, and calculation was therefore a comparatively slow process.

During the nineteenth century attempts were made to design automatic machines, that is, machines that would carry out the basic arithmetic processes automatically once the numbers to be processed had been entered into their registers. The first person to propose such a machine was the Cambridge mathematician, Charles Babbage.

Babbage called his proposed machine a 'difference engine'. Its main purpose was the calculation of various specialized mathematical tables, well-known at the time to be often very inaccurate. The machine was to have six registers which were so interconnected that it would be possible automatically to add the content of one register to another, and to transfer numbers between registers. A small prototype model was built in the early 1820s, and, as a result, Babbage received a substantial government grant to continue his work and produce a full-sized fully operative machine. This was never completed, and the project was eventually abandoned, largely because Babbage began to be inspired by the concept of a much more ambitious project. Babbage's ideas were, however, put to good effect by a Swiss engineer, Georg Scheutz, who constructed a working machine which was eventually used in the preparation of life tables.

Babbage's new inspiration was the creation of what he called an 'analytical engine'. Today, we would call such a machine a 'general-purpose computer'. As with his earlier project, the building of this was never completed. It was to consist of an intricate system of gears, rod, and linkages making up a store for retaining numbers, a processing unit for performing arithmetic operation, a control unit for regulating the correct sequence of operations, and input and output devices. The scale envisaged was much larger than anything which had been attempted before. For example, the store was intended to retain up to 1000 numbers, each up to 50 decimal places. The control unit was the nerve-centre of the whole contraption. It was this that effectively replaced the human operator essential to earlier machines.

One important feature of the proposed analytical engine was the use of punched cards. These had been invented in 1801 by the French loom-maker, Joseph Marie Jacquard, and were used for controlling the threads so that intricate patterns could be woven automatically. This principle has become a feature of the modern computer.

In 1890, an American statistician, Hermann Hollerith, introduced punched cards into a mechanical computer designed both to count and to sort which was used in the United States census. The machine is reputed to have saved the American government some two-million pounds, a considerable amount in those days even for a comparatively rich country. Hollerith's cards had to be punched by hand, thus the setting up of an operation on his machine was a slow process. This problem was partially solved some 26 years later by the invention of a small hand machine which would punch up to nine holes simultaneously.

Punched cards can be read automatically either by mechanical or by electrical means. The information received at any particular location on the card is of a yes–no character: either there is a hole or there is not. This is immediately suitable for handling logical information; we can record directly whether or not some particular object or property exists, whether a particular statement is true or false, whether an aperture is open or shut, and so on. However, the object of using the punched cards is primarily to handle numerical information, and in order to do this the information has to be translated into binary form.

Binary Arithmetic

Our decimal system requires precisely ten distinct numerals including a zero. We can write any whole number in the form

$$a_n a_{n-1} a_{n-2} \ldots a_1 a_0$$

where each a is any numeral 0 to 9 and we interpret the number represented as

$$(a_n \times 10^n) + (a_{n-1} \times 10^{n-1}) + (a_{n-2} \times 10^{n-2}) + \ldots + (a_1 \times 10) + a_0$$

Let us see what happens if we take our base to be 2 instead of 10. A number such as 1010 will now be interpreted as

$$(1 \times 2^3) \times (0 \times 2^2) + (1 \times 2) + 0$$

In other words 1010 expressed in base-two notation is the same as our 10 in decimal notation. What is crucially important here is that base-two notation requires only two numerals, 1 and 0. It is thus a binary numeral system, and this is precisely what is needed in a punched card machine.

In the binary system, the addition and multiplication tables are very simple indeed. They are just

$$0 + 0 = 0$$
$$0 + 1 = 1 + 0 = 1$$
$$1 + 1 = 10$$

and

$$0 \times 0 = 0$$
$$0 \times 1 = 1 \times 0 = 0$$
$$1 \times 1 = 1$$

(We should note here that binary 10 is 2 in the decimal system.)

Unfortunately, we have to pay a price for this simplicity: the number of places required to represent larger numbers increases much more rapidly than with the decimal system. For example, three binary places are needed to represent 4, four places to represent 8; 64 requires seven binary places, and 512 requires ten binary places.

The binary system can be extended to decimal fractions. Thus 0·1 represents 1×2^{-1} = decimal 0·5 (1/2); 0·01 represents 1×2^{-2} = decimal 0·25 (1/4); 0·001 represents 1×2^{-3} = decimal 0·125 (1/8), and so on.

The basic operations of arithmetic can be performed very easily. Decimal 129 + 87, for example, becomes

```
 10000001
  1010111
 --------
 11011000
```

Decimal 47·25 − 19·5 becomes

```
 101111·01
  10011·10
 ---------
  11011·11
```

Multiplication and division and even the extraction of square roots is just as simple. Thus decimal 35 × 23 becomes

```
    100011
     10111
----------
100011
  100011
   100011
    100011
----------
1100100101
```

and calculating the square root of decimal 126·5625 becomes

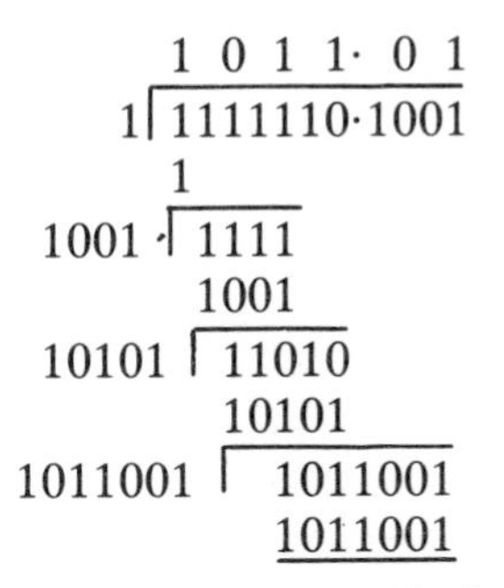

The calculated root 1011·01 is equivalent to decimal 11·25.

The binary numeral system should not be confused with counting by twos. Two-counting represents man's first attempts at building up a system of number words and is a part of our pre-history. Binary numerals were first suggested by Leibniz. His reasons were, curiously enough, theological. He believed that the two numerals required, 0 and 1, were symbolic of the creativity of God – God (represented by 1) created the universe out of nothing (represented by 0).

In order to make a machine perform binary arithmetic, all that is necessary is that mechanical or electrical linkages should perform according to binary addition and multiplication tables. Thus, with the punched card, the system needs to be so arranged that

· · · o and · · · o

produce

· · o ·

for addition, and

· · · o

for multiplication. A similar situation arises where we have currents through wires. If we have a simple switch which we can open and close, an open switch (that is, no current flowing)

can represent 0, and a closed switch (that is, current flowing)

can represent 1. Any number can thus be represented either on a series of parallel wires, or sequentially on a single wire rather like Morse Code. We do not have to be involved with the actual measurement of current with all the inaccuracies which can build up after a series of measurements – it is sufficient merely to be able to determine whether or not at any given moment there is a current.

Computers

We do not present any detailed discussion of computers. However, having taken the story of calculating machines as far as the Hollerith machine, we will bring it up to date with a brief survey of the main developments in the twentieth century.

Mechanical machines are by their very nature comparatively slow in performance. For speed of operation, electrification was essential. The first automatic general-purpose computer was built at Harvard University by the IBM Corporation, one of the largest manufacturers of punched-card machines. It was completed in 1944 and contained more than 500 miles of wire. By today's standards it was very slow, taking 3/10ths of a second for addition and 4 seconds for multiplication.

The first electronic machine appeared 2 years later. This was the ENIAC computer at the University of Pennsylvania. Valves replaced the moving parts, the switches, of the earlier computer and addition time was reduced to 1/5000th of a second. Since that time, speed of performance has been considerably improved and the invention of modern integrated circuits has meant that computers that once occupied the whole of a large room can now be housed in what is little more than a large piece of furniture. As modern techniques of microprocessing develop, it will be possible to have computers, having enormous capabilities, small enough to be inserted in a pocket.

The basic components of the modern computer have remained very much as Babbage conceived them. There is an input, a memory, a control, a processing unit, and an output. The operations to be carried out are controlled according to programs, which have to be prepared beforehand, though once prepared they can be stored in a computer's memory and called upon as and when required.

Preparing a program involves breaking down an overall operation into a sequence of simpler operations which the computer can carry out. Suppose that we wish to divide 1311 by 57. The computer may well be programmed to carry this out by a process of successive subtraction. The overall program, when written out, will look something like the following:

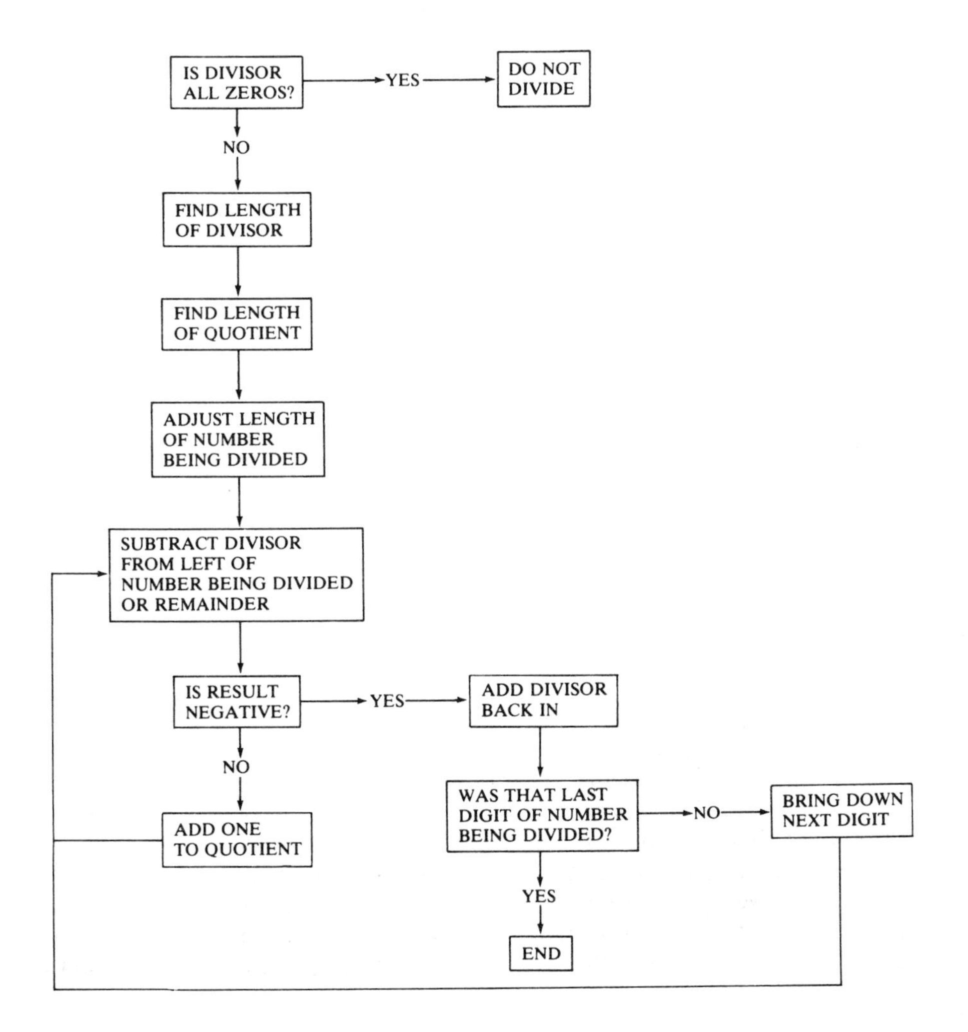

Assuming that, as is usually the case, the computer automatically converts decimal numbers to their binary equivalents we can now see just how this program deals with our division. We enter 1311 and 57 and instruct the computer to divide. The answer to the first question 'is divisor all zeros?' is 'no', so the operation proceeds. The length of the divisor is six binary digits, and the length of the quotient will be five digits. The dividend is now moved so that subtraction may start. We can represent what now begins (in decimal notation) as follows:

	→ 1	quotient;
57	1311	
	57	subtract;
	−44	negative remainder, so add 57;
	13	3 is not last digit, therefore
	131	bring down next digit;
	57	subtract;
	74	positive remainder so add 1 to quotient.

The subtraction is now repeated, and the calculation proceeds. Eventually, the answer to the question 'was that the last digit?' will be 'yes' and the computer will cease operation, printing out the final quotient as 23.

An important part of this operation was the repetition of certain cycles of operations. These are included in a program as a loop. One example is the loop:

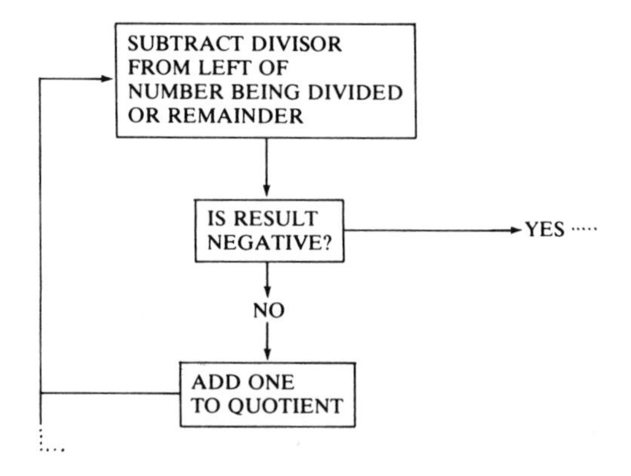

Programs have to be very carefully designed so that there is always a possible path out of a loop, otherwise the overall operation would never finish. The idea of a program loop, clearly a crucial concept in computer operation, goes back to the daughter of Lord Byron, the Countess of Lovelace. She was a friend of Babbage, and had acquired a thorough grasp of the principles underlying the analytical engine. She concentrated her attentions principally on what we now call programming and is said to have acquired at least as good a grasp of this as Babbage himself, and to be much better than he in explaining them to other people.

The possibilities inherent in the developments in computers which are now taking place are almost beyond comprehension. It is for this reason, that they are regarded with suspicion, if not fear, by many lay people.

Chapter Six
Unknown Numbers

Kits, cats, sacks and wives,
How many were going to St. Ives?

(Children's rhyme)

And wisely tell what hour o' th' day
The clock doth strike, by algebra.

(Samuel Butler)

In discussing methods of performing arithmetic calculations, we found that from time to time we had to make use of algebraic expressions to explain general points of principle and justify particular methods. Thus, the best way of stating that when we add two numbers together it does not matter which number we take first (the commutative property of addition) is to write down the simple expression

$$a + b = b + a$$

In describing such an expression as algebraic, what we usually imply is that we are using letters to stand for numbers. However, the real advantage here is that a comparatively long sentence is replaced by a short, clear and precise formula.

Until the nineteenth century, algebra was just the branch of mathematics in which symbols and numbers were manipulated together either to express general principles in a convenient form of shorthand or to solve equations. Such equations usually arose out of practical problems. The important point is that with very few exceptions the symbols stood for numerical quantities. This is still true today for many pupils in schools. They first meet algebra in the form of equations, the solutions of which are numbers. It is still true for many scientists and engineers, whose notebooks are full of algebraic expressions. Einstein expressed one of the greatest discoveries of all time, the special theory of relativity, as a simple equation

$$E = mc^2$$

From the nineteenth century onwards, algebra has come to have a more general meaning for mathematicians. The letters used in algebraic expressions no longer have to stand for numbers. Even familiar symbols such as + and ×

no longer necessarily denote addition and multiplication. Modern algebra has become the study of the manipulation of symbols in the abstract: it is no longer necessary to bear in mind what the symbols may or may not represent, what matters is that they have certain properties and that they combine with other symbols in certain ways.

We shall not discuss these modern developments because our concern is with numbers, and hence with algebra only in so far as it relates directly to numbers. This means that this chapter is about what we conventionally understand by the term 'equations'. The familiar unknown xs of algebra will, for us, always represent numbers.

Notation is crucial to algebra, yet it is surprising how much was known in ancient times about the solution of equations despite the fact that these had almost entirely to be written out in words. The absence of notation did mean, however, that there was a lack of generality in much of the algebraic thinking. General expressions were much more difficult to express in words than specific problems. Concentration centred on the actual mechanics of solutions rather than on the principle involved. Principle had very much to wait until suitable notation emerged.

The word 'equation' is derived from Latin *aequalis*. Essentially it means the act of balancing two quantities. Thus, in an equation such as

$$x^2 - 1 = 8$$

the left-hand side $x^2 - 1$ is balanced against the right-hand side 8. The word 'algebra' comes from the title of a book by al-Khwarizmi, *Hisâb al-jabr w'al-muqâbalah*, literally 'the science of reunion and opposition'. *Al-jabr* and *muqâbalah* were technical terms, referring to transferring negative terms to the other side of an equation and cancelling similar terms on either side of an equation respectively. Thus, in solving

$$x^2 - 1 = 8$$

we first transfer the negative -1 to the other side of the equation to give

$$x^2 = 9$$

This is *al-jabr*. As with his *Arithmetic*, al-Khwarizmi's work on algebra was translated into Latin and became influential in the West, and the word *al-jabr* passed into our vocabulary as 'algebra'. It also came to have the pseudo-medical meaning of 'bone-setter' and passed into Europe via Spain, where for a long time an *algebrista* meant a barber who performed bone-setting, and sometimes bleeding, as a sideline.

The Linear Unknown

Today, we would express the most general form of a linear equation as

$$ax + b = 0$$

where x is the unknown, and a and b are given numbers. The solution is obtained in two simple stages,

$$ax = -b$$

$$x = -\frac{b}{a}$$

In ancient times, there was no algebraic notation available and hence no absolutely general method of solution. Equations had to be written out in words, and hence each was considered very much as an individual problem, though it is clear from ancient documents that a number of strategies for solving equations were well understood. The Babylonians had a very simple strategy since both reciprocal and multiplication tables were available to them. Presented with a problem of the form $ax = b$, they looked up the reciprocal of a and its product with b, and so obtained the required answer.

A typical linear equation arises in a Babylonian problem concerning the differences in the amounts of money given to a family of brothers. It was (in our notation)

$$5x = 8$$

In the reciprocal tables, we find:

𒐊 𒌋𒐖

giving the reciprocal of 5 as $\frac{12}{60}$. In the table of multiples of eight, we find

𒌋𒐖 𒁹𒌍𒐋

The solution to the equation is therefore that the unknown is equal to $1\frac{36}{60}$.

Almost all the surviving Babylonian equations deal with practical problems of money, weight, volume or some other quantity involving units of mensuration. Since such units were sexagesimal, the tables used in calculations could be applied universally to any algebraic problem.

Most of the surviving Egyptian equations also deal with practical problems, though here we encounter different and less convenient methods of solution. A typical linear equation in the Rhind papyrus is expressed as:

Aha, its whole, its seventh,
it makes 19.

In modern notation we would write this as:

$$x + \tfrac{1}{7}x = 19$$

The word *aha*, once thought to be *hau*, meant 'heap', and was used to denote the unknown in a problem. The choice of this particular word underlines for us the original practical basis of Egyptian algebra.

We would solve this today by collecting the terms in x to give

$$\tfrac{8}{7}x = 19$$

thus obtaining the solution

$$x = 19 \times \tfrac{7}{8} = 16\tfrac{5}{8}$$

However, having no algebraic notation, the Egyptians were unable to collect terms and multiply and divide in this way. Their method of solution involved a principle which later became widely known in Europe as the *regula falsi*, the rule of false.

They began with the assumption that the unknown is equal to 7, a convenient choice because of the fraction $\frac{1}{7}$. This yields:

$$7 + 1 = 8$$

However, we now have 8 where we ought to have 19. They therefore resorted to a division of the kind described in Chapter 4. They wrote (in hieroglyphics):

8	1
16	2
4	$\frac{1}{2}$
2	$\frac{1}{4}$
1	$\frac{1}{8}$

This yields $2 + \frac{1}{4} + \frac{1}{8}$ which has to be multiplied by 7. So they now had to write out a calculation which transcribes as

	$2 + \frac{1}{4} + \frac{1}{8}$	1—
	$4 + \frac{1}{2} + \frac{1}{4}$	2—
	$9 + \frac{1}{2}$	4—
sum	$16 + \frac{1}{2} + \frac{1}{8}$	7

and thus eventually obtained the final solution, expressing $\frac{5}{8}$ in unit fractions.

Occasionally, they would choose unity as the initial assumption. This would often involve them in highly complicated calculations with unit fractions in which the scribes had to be highly skilled. Their results were usually checked by further calculations using the solution already obtained in place of the original assumption.

Also in the Rhind papyrus, we find a somewhat curious problem which appears to be some kind of inventóry. It translates:

Houses	7
Cats	49
Mice	343
Wheat	2401
Hekats	16807

We see that the numbers on the right are just successive powers of seven. However, when we recall the children's rhyme about the man going to St. Ives, it is tempting to see its historical origins in this Egyptian problem. Since it also appears in a similar form in Leonardo of Pisa's *Liber abaci*, our suspicion that it may indeed be several thousand years old is strengthened. Leonardo's version does not, however, have the cunning twist of the modern rhyme, nor does the Egyptian version.

Although the purpose of developing mathematical skills in Egypt was to use them in practical problems, many of the calculations in surviving papyri are not related to any specific application – they are just exercises in manipulation such as we find in schools today. However, despite the skill achieved with unit fractions, the solution of equations in Egypt was, by comparison with Babylonian methods, so complicated that it is small wonder that the latter made far greater progress in algebra.

Greek mathematicians could solve linear equations without difficulty. Their interests, however, lay very largely with geometry and number theory, and thus the application of elementary algebra to everyday problems was scarcely regarded as a reputable mathematical activity, though it was taught in the schools. Occasionally we find a linear equation concealed in the form of a riddle, as with the description of the life of Diophantus (third century A.D.) found in a collection of problems dating from some two centuries later. The riddle reads:

> God granted him youth for the sixth part of his life, and after a twelfth part He clothed his cheeks with down. After a seventh part, He gave him the light of marriage, and five years later granted him a son, whom Fate claimed after he attained the measure of half his father's life. He consoled himself with the science of numbers for four more years, and then his life ended.

If we let x represent the span of Diophantus' life, we obtain the linear equation:

$$x = \tfrac{1}{6}x + \tfrac{1}{12}x + \tfrac{1}{7}x + 5 + \tfrac{1}{2}x + 4$$

which tells us, after a little manipulation, that Diophantus lived to be 84. Diophantus himself would hardly have thought such a problem worthy of his attention.

Algebra in India can be traced back to some seven or eight centuries before Christ. The earliest equations seem to have been mainly geometrical in origin. For example, in the *Sulba*, we find a problem which requires us to find the length of the side of a rectangle whose other side is given and whose area is equal to that of a given square. This is just equivalent to the linear equation

$$ax = b^2$$

Like the Egyptians, the Hindus made considerable use of the rule of false, and we find examples dating from the beginning of the first century A.D. Given a problem leading to an equation of the form

$$ax + b = c$$

they knew that making an assumption g such that

$$ag + b = d$$

enabled them to calculate the true value of the unknown x by calculating according to the formula

$$x = \frac{c - d}{a} + g$$

They could not, of course, express this formula symbolically as we can do, but they were nevertheless able to carry out the calculations which it requires.

The rule of false disappears from the Indian scene from about the sixth century onwards, presumably because the introduction of limited forms of algebraic notation enabled more direct methods of solution to be undertaken. Brahmagupta is quite explicit about how to solve an equation of the form

$$ax + b = cx + d$$

He writes:

> In the case of an equation with one unknown, take the difference of the known terms in reverse order divided by the difference of the coefficients of the unknown.

In other words, the solution is just

$$x = \frac{d - b}{a - c}$$

Indian methods of solution, including the earlier rule of false, passed to the Arabs and through them to the West. Along with the rule of three, the rule of false was extensively explained in European texts from the time when it first appeared in a Latin translation of al-Khwarizmi's *Algebra*. Indeed, it was extended to a rule of double false, involving two different assumptions of the value of the unknown. To see how this method works, we will again return to the general form of a linear equation,

$$ax + b = 0$$

Suppose that our two assumptions are g_1 and g_2 from which we obtain two false results

$$ag_1 + b = f_1$$
$$ag_2 + b = f_2$$

The results f_1 and f_2 are, of course, false if they are not equal to 0. If for example, a problem leads to the equation

$$5x - 10 = 0$$

we can make assumptions $x = g_1 = 3$ and $x = g_2 = 1$, which give us $f_1 = 5$ and $f_2 = -5$. Returning to the general case, we obtain by subtraction

$$a(g_1 - g_2) = f_1 - f_2$$

We now eliminate a from the pair of equations giving us f_1 and f_2 by multiplying the first by g_2, the second by g_1, and then subtracting, that is:

$$ag_1g_2 + bg_2 = f_1g_2$$
$$ag_1g_2 + bg_1 = f_2g_1$$

from which we obtain

$$b(g_2 - g_1) = f_1g_2 - f_2g_1$$

We now divide by each side of the equation

$$a(g_1 - g_2) = f_1 - f_2$$

which gives us:

$$-\frac{b}{a} = \frac{f_1 g_2 - f_2 g_1}{f_1 - f_2}$$

This is precisely the formula we want. Thus, in our example, we obtain the solution for x by substituting for f_1, f_2, g_1 and g_2:

$$x = \frac{(5 \times 1) - (-5 \times 3)}{5 - (-5)}$$

$$= \frac{20}{10}$$

$$= 2$$

If we compare this with our modern approach where x is given directly as $-b/a$, we may well wonder how the rule of double false became so popular, but we must remember that so long as problems were stated very largely in words and there was no satisfactory notation by which an equation could be expressed symbolically, the clumsiness of the earlier methods could not be easily exposed.

The principles of the rule of double false had been known to the Arab mathematicians, who presented it in a slightly modified form known as *Alm bi'l kaffatain*, the 'method of the scales'. This name came from a simple diagram which enabled the solution to be written down very quickly:

The two lines on the left represent f_1 and f_2 and the two on the right g_1 and g_2. The cross in the centre indicates how they are to be multiplied together. For our example,

$$5x - 10 = 0$$

where $f_1 = 5$, $f_2 = -5$, $g_1 = 3$ and $g_2 = 1$, we have

This assists us in seeing that we have to calculate

$$\frac{(5 \times 1) - (-5 \times 3)}{5 - (-5)}$$

to find x.

We have so far confined ourselves to linear equations, where there is just one unknown. We shall be considering linear equations with more than one unknown later. This particular classification of equations, so useful for our

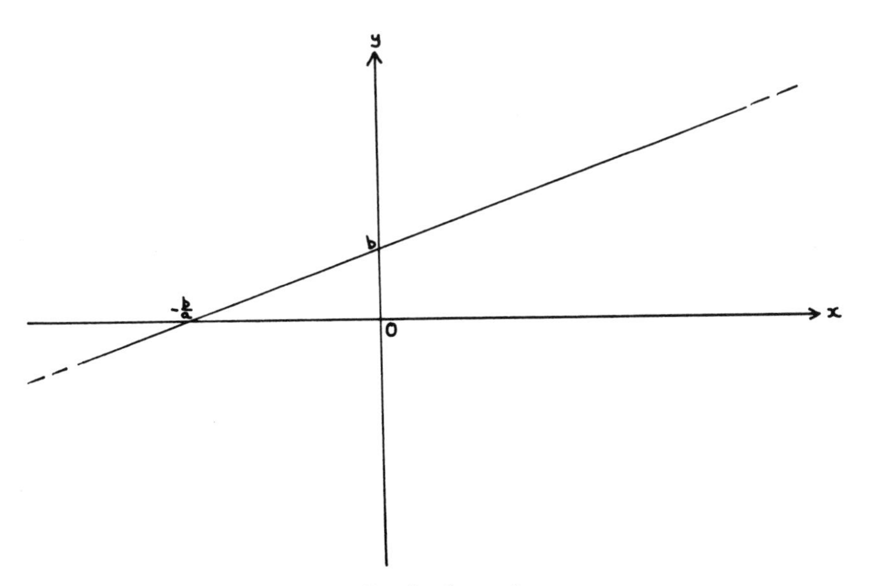

Graph of $ax + b$

purposes here, is of comparatively modern origin, dating from about the beginning of the seventeenth century. Earlier Indian and Arab classifications had tended to call equations simple or compound according to the number of terms involved on each side of the equality. If either side involved addition or subtraction of two or more terms, the problem would be classified as compound; otherwise it was simple.

The Greeks, however, with their geometrical approach, classified problems as being either plane, solid, or linear. Plane problems included those related both to the straight line and the circle. What we now call a linear problem is one which can be represented by a straight line graph. Thus if we plot the value of $ax + b$ against different values of x, we get a straight line of slope a cutting the vertical axis at b and the horizontal axis at $-b/a$, the value of x for which $ax + b$ is zero and hence the solution of the equation

$$ax + b = 0$$

The problems which the Greeks classified as linear were the most complex of all, and involved curves of greater complexity than the circle and the conic sections.

The Square of the Unknown

The most general form of a quadratic equation can be expressed

$$ax^2 + bx + c = 0$$

The unknown and its square both appear, though in simpler equations the term bx may not be included. (We should note that we can always reduce the coefficient of the square of the unknown to unity by dividing through by a.) Such equations are 'quadratic' because the highest power of the unknown occurring is its square. There will be two solutions, though these will not necessarily be real numbers. These solutions can be obtained directly by using the formula

$$x = \frac{-b \pm \sqrt{b^2 - 4ac}}{2a}$$

It is because we have a good notation for algebra that we can express the general equation and its solutions in this way. Until such notation became available in the seventeenth century, quadratic equations were classified into different kinds according to which terms were present and whether the given coefficients were positive or negative. Special solution methods were then used for each class of equation.

We find that the solution of quadratic equations was well-developed in Babylonian algebra. This is in contrast to the situation in Egypt, where we do find occasional quadratic problems being mentioned, for example in the Berlin and Kahun papyri, but it does not seem that any systematic treatment of these was known. We must, however, remember that, compared with what has survived of Babylonian mathematics, the few surviving Egyptian mathematical papyri represent only very scant evidence of Egyptian mathematical achievement.

Both the Egyptians and the Babylonians calculated tables of square roots, approximations being used in many cases so that the problem of irrational numbers, such as $\sqrt{2}$, was not appreciated. Presumably, it was thought that the representation of these by a sequence of unit or sexagesimal fractions would eventually come to an end if carried far enough. The Babylonians had an approximation procedure for calculating the square roots of numbers which are not exact squares. Their method was based on the relation

$$\sqrt{n^2 + e} \simeq n + \frac{e}{2n}$$

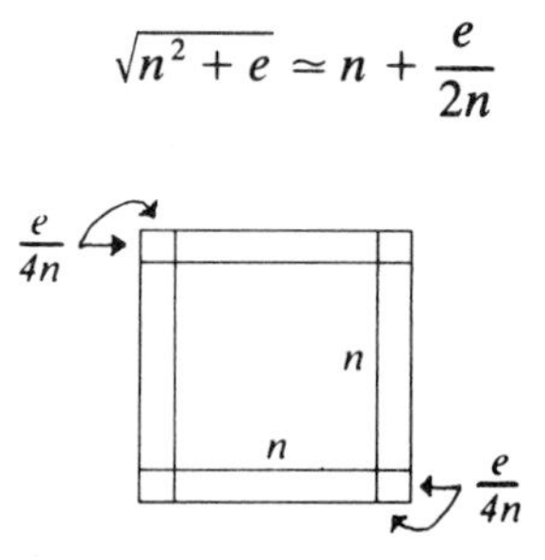

It is possible to deduce this quite simply from the geometrical diagram shown. The approximation simply involves neglecting the four small squares at the corners. We do not know if the Babylonians used such geometrical means of justifying their methods. They were, however, typical of Greek geometrical algebra, and may have been known earlier.

The square root tables enabled equations of the form

$$x^2 = c$$

to be solved immediately, though only the positive root would be recognised.

We find equations of the forms

$$x^2 + bx = c$$

$$x^2 = bx + c$$

and

$$x^2 + c = bx$$

occurring frequently in Babylonian tablets. Often the unknown was expressed explicitly as a length, and its square as an area. This did not, however, prevent the subtraction of one from the other. For example, we find the equation

$$x^2 - x = 870$$

expressed as

I have subtracted the side of a square from its area; result 14,30.

(14,30 is the sexagesimal equivalent of decimal 870.) Calculation of the solution of this equation is explained in detail. We are first to take half of the coefficient of the side, the negative being ignored, so that we have half of unity – in sexagesimal notation ;30. We must now square this, obtaining (from tables) ;15 that is, $\frac{15}{60}$ or $\frac{1}{4}$. We add this to 14,30 so as to obtain 14,30;15. From tables, we find the square root of this to be 29;30, and we add ;30 to this, the final solution being 30.

Let us now see just what we have done in this calculation. If we compare

$$x^2 - x = 870$$

with the general form of the quadratic equation,

$$ax^2 + bx + c = 0$$

we find that $a = 1, b = -1, c = -870$. Our modern formula

$$x = \frac{-b \pm \sqrt{b^2 - 4ac}}{2a}$$

can be slightly rearranged as

$$x = -\frac{b}{2a} \pm \sqrt{\left(\frac{b}{2a}\right)^2 - \frac{c}{a}}$$

by separating the two parts of the numerator and including the denominator under the square root sign. If we now substitute for a, b and c we have

$$x = \tfrac{1}{2} \pm \sqrt{(\tfrac{1}{2})^2 + 870}$$

and this follows the Babylonian solution precisely, except that only the positive square root is taken. The second solution, $x = 29$, is not obtained.

Quite often, in order to solve an equation of the form

$$x^2 + c = bx$$

the Babylonians would use a method involving the solution of a pair of equations

$$\left.\begin{aligned} x + y &= b \\ xy &= c \end{aligned}\right\}$$

These simultaneous equations can be seen to be equivalent to the original quadratic equation if we substitute from the second equation into the first so as to eliminate y. We then get

$$x + \frac{c}{x} = b$$

and, on multiplying through by x,

$$x^2 + c = bx$$

Similarly, an equation of the form

$$x^2 - bx = c$$

would be solved using the pair of equivalent equations

$$\left.\begin{aligned} x - y &= b \\ xy &= c \end{aligned}\right\}$$

A special case arose with a problem found in early Babylonian tablets, which we can translate as

A number and its reciprocal are equal to a number.

In our notation, this is just

$$x + \frac{1}{x} = b$$

or, multiplying by x,

$$x^2 - bx = 1$$

The solution method adopted was quite straightforward, and is similar to the earlier example which we discussed. Calculate $b/2$, then $(b/2)^2$, subtract 1, look up the square root in tables, and add this to $b/2$. In other words, they calculated

$$\frac{b}{2} + \sqrt{\left(\frac{b}{2}\right)^2 - 1}$$

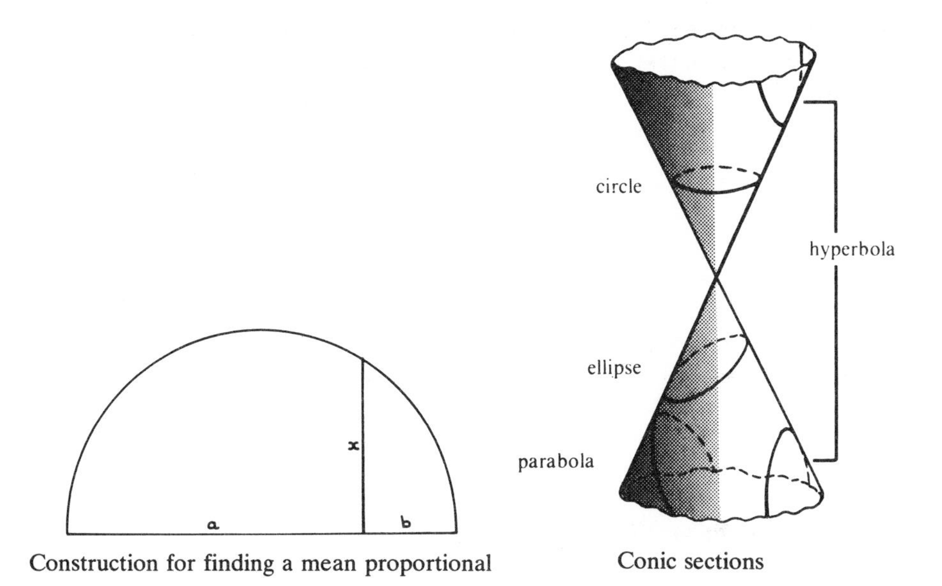

Construction for finding a mean proportional

Conic sections

which is again equivalent to just another application of our modern general formula.

Greek algebra was for the most part highly geometrical and was mainly concerned with quadratic equations. These arose out of what the Greeks called 'finding a mean proportional', out of numerous problems associated with areas, and from the conic sections. The latter are formed when a cone is sliced in various different ways.

A mean proportional between two numbers a and b is a number x such that

$$a:x = x:b$$

If we write this as

$$\frac{a}{x} = \frac{x}{b}$$

and cross-multiply, we see that this is just the equation

$$x^2 = ab$$

The Greeks solved this problem by a geometric method which involved constructing a semi-circle on a length $a + b$, and they used geometric reasoning to justify this method. Indeed, with this idea of geometric proof, we find a great change from the earlier Babylonian and Egyptian algebra, where specific solutions were justified by re-working equations using the calculated answer. Although general methods were known, they were very seldom described and we have no evidence that they were justified by appeal to general proofs, geometric or otherwise.

The Greeks resorted to geometric methods of construction very largely because their theory of numbers, being limited to numbers which can be expressed in terms of relations between whole numbers, could not cope with irrational quantities. A quantity such as $\sqrt{2}$, the mean proportional between 1 and 2 and the length of the diagonal of a unit square, could only be constructed geometrically; it could not be calculated. Indeed, it was not regarded by the Greeks as a number.

We find the discussion of the general rules for solving quadratic equations by geometric methods in Book VI of Euclid's *Elements*, which also owes much to earlier discussions in Book II. Propositions 28 and 29 of Book VI are as follows:

> *Proposition 28:* To a given straight line to apply a parallelogram equal to a given rectilinear figure, and deficient by a parallelogrammic figure similar to a given one; thus the given rectilineal figure must not be greater than the parallelogram described on the half of the straight line and similar to the defect.

> *Proposition 29:* To a given straight line to apply a parallelogram equal to a given rectilineal figure and exceeding by a parallelogrammic figure similar to a given one.

At first sight, these two propositions do not seem to have anything to do with algebra. However if we analyse what we are being asked to do, we find that we are in fact solving the quadratic equations

$$x^2 - bx + c^2 = 0$$

and

$$x^2 - bx - c^2 = 0$$

To see this, it is easier to consider the special case where the parallelograms are all rectangles and the given parallelogram is a square.

Let AB be the given straight line. We are asked in each case to 'apply a parallelogram' to it equal to the given area. This just means that in our special cases we have to find a point P on AB or AB extended and erect a rectangle $APQR$ satisfying the given conditions on AP. The 'parallelogrammatic figure' by which it is 'deficient' in one case and 'exceeding' in the other is just the

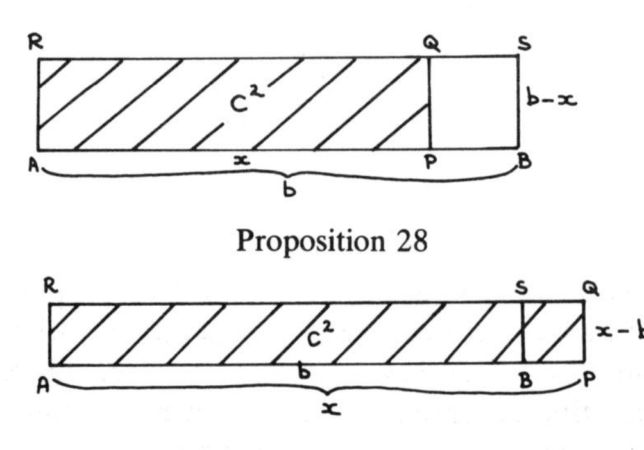

Proposition 28

Proposition 29

square $PQSB$. If the length of AB is b, the length of AP is x, and the given area is c^2, the areas which must be equal to c^2 are $x(b - x)$ in the case of Proposition 28 and $x(x - b)$ in the case of Proposition 29. The two equations above then follow directly.

We used c^2 rather than c in our equations, when discussing Greek geometric algebra, because the geometric approach allows us to add and subtract areas and add and subtract lengths, but does not allow us to add a length to an area. Thus all the terms in an equation such as

$$x^2 - bx + c^2 = 0$$

are considered to be areas. It would be inappropriate to write

$$x^2 - bx + c = 0$$

The later Greek mathematicians of the Alexandrine period abandoned the classical geometrical approach and, in effect, reverted to Babylonian methods. This meant that they saw nothing unacceptable in adding a term in x to a term in x^2. We find, for example, in the work of Heron (around A.D. 100), the problem

The sum of the area and the perimeter of a square is 896; find its side.

This is just the quadratic equation

$$x^2 + 4x = 896$$

Heron's method is to add 4 to each side, thus making each side a perfect square, and then to take square roots. Thus,

$$x^2 + 4x + 4 = 900$$

that is,

$$(x + 2)^2 = 30^2$$

whence the solution $x = 28$. This is, of course, explained in words, and is our modern method of completing the square. However, Heron offers no proof of his method – another departure from the classical period.

Diophantus, the greatest of all Greek algebraists, distinguished three kinds of quadratic equation:

$$ax^2 + bx = c$$

$$ax^2 = bx + c$$

and

$$ax^2 + c = bx$$

Each of these had its own particular method of solution, equivalent to interpretations of our modern formula, though only positive a, b, c and positive roots were admitted since negative numbers were not recognized. Negative numbers were first accepted in a limited way in India around the sixth century, by which time the Greek developments had long ceased.

In the most ancient Indian documents, those from the Vedic period (before 1000 B.C.), we find accounts of the construction of altars which involve the solution of quadratic equations of the form

$$ax^2 + bx = c$$

Later works, such as the Bakshâlî treatise (around A.D. 200), clearly show familiarity with certain types of quadratic equation. By the sixth century many examples were being discussed in detail, often related to problems of interest on capital. Aryabhaṭa gives the rule for solving the following problem:

> A sum of 100 is lent for a month. The interest is then lent for 6 months. After this period the original interest together with the interest on the interest amounts to 16. Find the original rate of interest per month.

This requires us to solve the quadratic equation

$$6x^2 + 100x - 1600 = 0$$

since, if x is the percentage rate of interest per month, the first interest is just x and the interest on this interest over 6 months is $6x^2/100$. The sum of these is 16. The solution is stated to be as follows:

> Multiply the sum of the interest on the principal and the interest on the interest by the time and by the principal. Add to this result the square of half the principal. Take the square root of this. Subtract half the principal and divide the remainder by the time. The result will be the interest on the principal.

If we do precisely this, we calculate

$$\frac{\sqrt{(16 \times 100 \times 6) + \left(\frac{100}{2}\right)^2} - \frac{100}{2}}{6}$$

If we apply our modern formula, we would calculate

$$\frac{-100 \pm \sqrt{100^2 + (4 \times 6 \times 1600)}}{12}$$

The two calculations are effectively the same except that the negative root, which does not give a practical solution of the problem, is not considered.

Brahmagupta gives two rules for the solution of quadratic equations. The first of these reads:

> The absolute quantities multiplied by four times the coefficient of the square of the unknown are increased by the square of the coefficient of the middle term; the square root of the result being diminished by the coefficient of the middle term and divided by twice the coefficient of the square of the unknown, is the value required.

In other words, this is our

$$x = \frac{-b + \sqrt{b^2 - 4ac}}{2a}$$

The second rule gives the variation equivalent to

$$x = \frac{\sqrt{\left(\frac{b}{2}\right)^2 - ac} - \frac{b}{2}}{a}$$

Later writers, such as Śrîdhara (eighth century) and also Bhâskara (twelfth century), on whom we have to rely for an account of much of the content of Śrîdhara's works, show that they are well aware that a quadratic equation usually has two roots. For example, Bhâskara writes:

> The square of a positive number, and of a negative number also, is positive; and the square root of a positive number is twofold, positive and negative. There is no square root of a negative number, for no negative number is a square.

He does point out later, however, that though taking two roots is theoretically correct, one of them may sometimes lead to an impractical solution and should then be rejected.

Knowledge of Indian algebra passed northward into China and westwards to the Arabs. However, the Arabs were also collectors and translators of Greek works and hence their algebra includes a number of the Greek geometrical proofs of solutions of equations. A certain limited amount of algebraic notation had been introduced both in India and during the later period of Greek mathematics. This, however, was not adopted by the Arab mathematicians, who returned to the somewhat cumbersome situation where equations had to be written in words. They also failed to adopt the negative numbers of the Hindus.

Al-Khwarizmi distinguished three types of quadratic equation. We would write these today as

$$x^2 + bx = c$$

$$x^2 + c = bx$$

and

$$x^2 = bx + c$$

In each case the unknown was referred to by a word which translates as 'thing' and the square of the unknown by a word which translates as 'power'. The first of these equations, therefore, appears as:

> Power and things equal to a number.

The equations are always stated in such a way that no negative term is

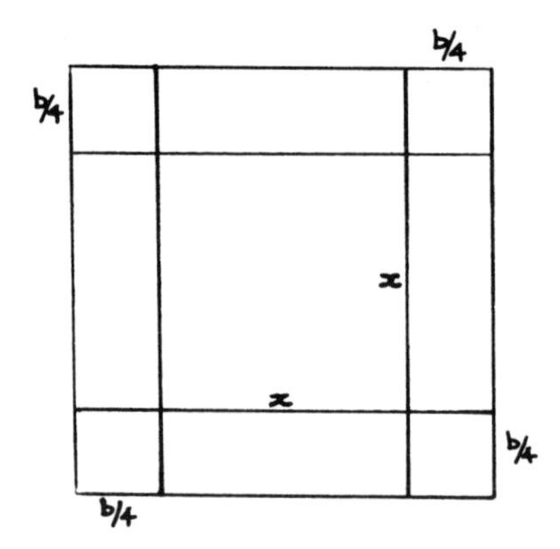

included, and, of course, there was no question of negative roots being accepted.

Proofs of the methods given were geometrical. For example, the method for the solution of an equation of the form

$$x^2 + bx = c$$

was equivalent to the application of the formula

$$x = \sqrt{c + \frac{b^2}{4}} - \frac{b}{2}$$

This method was justified by an appeal to the figure shown above. The area of the large square can be expressed in two ways, either as

$$\left[x + 2\left(\frac{b}{4}\right)\right]^2$$

or as

$$x^2 + 4\left(\frac{bx}{4}\right) + 4\left(\frac{b^2}{16}\right)$$

We obtain the former by taking the total length of one side and squaring it, and the latter by adding the areas of the five squares and four rectangles into which it can be divided. If we now simplify these two expressions and write them as equal, we have

$$\left(x + \frac{b}{2}\right)^2 = x^2 + bx + \frac{b^2}{4}$$

But, going back to the original equation, we see that the right-hand side is just $c + (b^2/4)$. Taking square roots, we have

$$x + \frac{b}{2} = \sqrt{c + \frac{b^2}{4}}$$

and the formula follows immediately by subtracting $b/2$ from each side. This is a typically Greek approach, and was clearly adopted from Greek works which had been translated into Arabic. Other Arab mathematicians such as Al-Karkhi and Omar Khayyam (of Rubáiyát fame) had slightly different forms of the formulae for the three types of quadratic equation, but their approach followed the pattern set by al-Khwarizmi. Equations of the form

$$x^2 + bx + c = 0$$

which have negative solutions, were not considered.

Arabic algebra was first introduced into the West by Leonardo of Pisa. In the *Liber abaci*, he distinguished between simple and compound forms of quadratic equation, the former including the three types:

$$ax^2 = bx$$

$$ax^2 = c$$

and

$$bx = c$$

and the latter the three types:

$$ax^2 + bx = c$$

$$bx + c = ax^2$$

and

$$ax^2 + c = bx$$

These were again carefully written out so as to avoid negative numbers. He always began by dividing through by a so as to make the coefficient of the power unity. He did notice that, for the equation

$$ax^2 + c = bx$$

it was possible to obtain a second positive solution by taking the negative root in the formula, provided, of course, that the term under the square root sign is not negative. Virtually no further advances were made until the sixteenth century.

Michael Stifel was the first to break the established convention and use negatives in his equations. He considered the three kinds of quadratic equation which we would write as

$$x^2 = c - bx$$

$$x^2 = bx - c$$

and

$$x^2 = bx + c$$

and gave one rule for solution, equivalent to the formula

$$x = \sqrt{\left(\frac{b}{2}\right)^2 \pm c} \pm \frac{b}{2}$$

The coefficients were still all taken to be positive. The positive or negative choices in the formula were taken to accord with the way in which the original problem had been stated.

Rafael Bombelli reintroduced the method of completing the square. He discusses, for example, the equation we write as

$$3x^2 + 6x = 24$$

He first multiplies through by 3 to make the power term a perfect square. This gives

$$9x^2 + 18x = 72$$

He now completes the square on the left-hand side by adding 9 to both sides:

$$9x^2 + 18x + 9 = 81$$

that is,

$$(3x + 3)^2 = 81$$

Taking positive roots, he obtains

$$3x + 3 = 9$$

and, hence,

$$x = 2$$

He does, however, on occasion find two solutions. For example, he discusses the equation which we write as

$$x^2 + 12 = 8x$$

Expressing it first in the form

$$x^2 - 8x + 12 = 0$$

he completes the square, by adding 4 to each side, obtaining

$$(x - 4)^2 = 4$$

and, hence,

$$x = 2$$

However, he now returns to the completed square and reverses the terms, so that we have

$$(4 - x)^2 = 4$$

from which

$$x = 6$$

Neither of these solutions follows the rule which Bombelli actually advocated in his text.

A real advance occurred with Francois Viète. Not only did he introduce a systematic algebraic notation but he also replaced methods based on geometric proofs by strictly algebraic ones. For example, in order to solve an equation of the form

$$x^2 + bx = c$$

he assumed that x could be expressed as the sum of two numbers u and z. Substituting into the equations gives

$$(u + z)^2 + b(u + z) = c$$

or

$$u^2 + (2z + b)u + z^2 + bz = c$$

If we now choose z to be equal to $-b/2$, we can remove the term in u and so obtain

$$u^2 - \frac{b^2}{4} = c$$

giving

$$u = \sqrt{c + \frac{b^2}{4}}$$

Since we have already chosen the value of z, we arrive at the solution of the equation by adding the calculated value of u. This method is just another example which is equivalent to using our modern formula, except that only the positive root in the expression for u was allowed.

Viète studied the relationships between the coefficients and roots of quadratic equations and also introduced an approximation method. He recognized, for example, that an equation which can be expressed in the form

$$x(u + z) - x^2 = uz$$

has solutions $x = u$ and $x = z$. His approximation method required him first to obtain a root close to the correct value. The true root could then be expressed as the approximate root plus a comparatively small quantity. This is again a replacement of x by $u + z$, but in this case u is known to be close to x and hence z must be small by comparison. By substituting into the original equation and dropping the z^2 term (on the grounds that the square of a small quantity is negligible), we obtain a formula for x which gives an even better approximation than u. We can now start all over again, taking u to be this improved approximation, and repeat the process. This method was very general and Viète applied it to equations of higher order. However, the true root was often reached so slowly that the amount of calculation involved in carrying out many successive applications of the approximation formula was described as 'work unfit for a Christian'.

We have seen how solutions of quadratic equations by formula and by completing the square had been known from early times. The other method which is widely taught in schools today is solution by factorization. The first serious attempts at using this method appear in the work of Thomas Harriot (seventeenth century), an English mathematician whose achievements have often been under-rated.

Today, we are accustomed to the idea that every quadratic equation has two roots, which may be a pair of real numbers or a pair of complex numbers – numbers of the form $a + b\sqrt{-1}$, usually written as $a + ib$. Thus if these roots are m and n, we have

$$\left.\begin{aligned} x &= m \\ x &= n \end{aligned}\right\}$$

or

$$\left.\begin{aligned} x - m &= 0 \\ x - n &= 0 \end{aligned}\right\}$$

and, multiplying these together, we have

$$(x - m)(x - n) = 0$$

This is just the quadratic equation

$$x^2 - (m + n)x + mn = 0$$

that is, a general quadratic of the form

$$x^2 + bx + c = 0$$

where $b = -(m + n)$ and $c = mn$. Thus, to solve such a quadratic equation by factors, we need to find by inspection two numbers m and n whose sum is $-b$ and whose product is c. For example, given the equation

$$x^2 + x - 6 = 0$$

we look for two numbers whose sum is -1 and whose product is -6. We can easily see that these numbers must be 2 and -3. These are therefore the required roots of the equation, which factorizes as

$$(x - 2)(x + 3) = 0$$

We can find factors by inspection without necessarily reducing the coefficient of x^2 to unity. Consider the equation

$$2x^2 - 11x + 5 = 0$$

To avoid working with fractions, we look for two factors which multiplied together give $2 \times 5 = 10$ and which added give 11. These are obviously 10 and 1. The equation will therefore factorize as

$$(2x - 1)(x - 5) = 0$$

giving $x = \frac{1}{2}$ and $x = 5$ as the two solutions. In other words, for the general equation

$$ax^2 + bx + c = 0$$

we look for two numbers whose product is ac and whose sum or difference (accordingly as c is positive or negative) is b. We then arrange the factors appropriately. This method, although popular in teaching, is not generally applicable because it relies on equations having factors which can easily be recognized. Most equations arising in practical applications do not turn out to be so convenient, and engineers are more likely either to make use of the general formula or to use an approximation method.

The proof of a general theorem, stating that every equation of degree n has n roots and hence that every quadratic equation has two roots, had to wait until the eighteenth century and, of course, included complex roots. Although complex roots must occur in pairs and hence quadratic equations are the first which can actually have such roots, they arose initially out of the study of cubic equations, that is, equations in which the highest power of the unknown x occurring is x^3.

If we plot the value of $ax^2 + bx + c$ against values of x we shall obtain a graph which is one of the conic sections – a parabola. It will cut the vertical axis at c (obtained by putting x equal to 0) and the horizontal axis at the two values of x which are the solutions of the equation:

$$ax^2 + bx + c = 0$$

If we have such a parabola which does not cut the x-axis, as is the case with $x^2 + c$, c positive, the roots of the equation $x^2 + c = 0$ must be complex. Indeed, in this case we just have

$$x^2 = -c$$

from which

$$x = \sqrt{-c}$$

$$= \pm i\sqrt{c}$$

Occasionally, we have the special case where the parabola touches rather than crosses the horizontal axis. In this case the two roots of the corresponding equation coincide, that is, the equation is of the form

$$(x - m)^2 = 0$$

Theoretically, there are still two factors and hence two roots, the two factors being $(x - m)$ and $(x - m)$ giving equal roots m and m. Multiplying out, we have

$$x^2 - 2mx + m^2 = 0$$

If we now substitute the coefficients into our general formula

$$x = \frac{-b \pm \sqrt{b^2 - 4ac}}{2a}$$

we see that

$$b^2 = 4ac = 4m^2$$

and hence the term under the square roots is equal to 0. This removes that part of the solution from which two distinct roots can be obtained.

The Greeks were well aware of the parabola and its properties, as of the other conic sections. However, their approach was to consider loci, that is, they were interested in the relationship between the conic sections and expressions relating two variables x and y, and not in the graphical solution of equations. Loci were again extensively studied in the seventeenth century, especially by Descartes. It is very largely due to Descartes that we owe both the modern approach to the classification of equations by degree and the re-marriage of geometry and algebra which made the graphical solution of equations possible.

Higher Powers of the Unknown

The equation of the next degree, the cubic equation, includes the unknown raised to the third power. The most general form of such an equation is

$$ax^3 + bx^2 + cx + d = 0$$

We find examples of cubic equations in Babylonian tablets. As with quadratic equations, these were solved using tables of values of n^3 and of $n(n^2 + 1)$. Solution of the cubic equations

$$x^3 = d$$

and

$$x^2(x + 1) = d$$

could be read from the tables immediately. Thus, the problem of finding the side of the cube, twelve times whose volume is equal to 1;30 (our $1\frac{1}{2}$) is solved by first looking up in a table of multiples of ;5 (our $\frac{1}{12}$) the product of ;5 and 1;30 and then looking in a table of cubes to see what number when cubed was equal to this product. The correct solution ;30 (our $\frac{1}{2}$) was thus obtained very easily.

Another Babylonian problem leads to an equation which we would write as

$$x^2(12x + 1) = \frac{7}{4}$$

The method of solution involved first multiplying both sides by $12^2 = 144$ which gives

$$(12x)^2(12x + 1) = 252$$

and then, treating $12x$ as n, looking in $n^2(n + 1)$ tables to find what n gives 4,12 (our 252). The tables give $n = 6$, hence $12x = 6$ and the solution $x = ;30$ (our $\frac{1}{2}$) follows immediately. Thus, any equation of the form

$$x^2(ax + 1) = d$$

could be solved.

One of the three most famous problems in Ancient Greece was that known as the 'Delian problem' or the 'duplication of the cube'. Along with attempts to square the circle and trisect an angle, this problem engaged the best of Greek mathematicians over many centuries. Hippocrates of Chios reduced the problem, essentially that of finding the side of a cube whose volume is twice that of a given cube, to the problem of finding two mean proportionals between v and $2v$, that is, finding x and y such that

$$v:x = x:y = y:2v$$

From

$$v:x = x:y$$

we get

$$y = \frac{x^2}{v}$$

and from

$$x:y = y:2v$$

we get

$$x^3 = 2v^2$$

after substituting for y. According to the principles of Greek mathematics this had to be solved by a geometric construction using straight edge and compasses only.

Various solutions were proposed of this insoluble problem, not finally proved to be so until the nineteenth century. Much good mathematics came out of the work on it. Some of the proposed solutions were mechanical, involving sliding rules; others provided approximate solutions only. Geometric solutions which gave a correct theoretical answer, of course, failed to comply with the limitation to straight-edge and compasses. These various kinds of solution are described for us by Pappus and Eutocius.

We find a problem in the work of Diophantus which is equivalent to solving the cubic equation:

$$x^3 + x = 4x^2 + 4$$

Unfortunately, the method of solution is not stated, though the solution given, 4, is correct. It seems likely that this solution was obtained simply by dividing both sides by $x^2 + 1$.

Very little was achieved with cubic or higher order equations by either the Hindus or Arabs, though one or two special cases were considered and geometrical solutions involving conic sections on Greek lines were quoted by Arab mathematicians. Bhâskara did discuss the possibility of reducing the solution of cubic equations to solutions of linear and quadratic equations by making appropriate substitutions, but he was unable to find any general method of doing this. He does, however, solve the following problem successfully:

> What is the number which, on being multiplied by twelve and increased by its cube, is equal to six times its square added to thirty-five?

This is just the cubic equation

$$12x + x^3 = 6x^2 + 35$$

The solution is obtained by first collecting all the terms in the unknown on the left-hand side:

$$x^3 - 6x^2 + 12x = 35$$

and then subtracting 8 from both sides. This gives

$$(x - 2)^3 = 27$$

from which

$$x - 2 = 3$$

is immediately obtained, and the solution, 5, follows.

Omar Khayyam studied various types of cubic equation, and gave a list of those which have positive roots. His approach was mainly geometrical and involved finding intersections of conics. Although solutions of some cubic equations were obtained algebraically, these were all special cases in which either terms of the general cubic were missing or the coefficients were such that a solution was clear virtually by inspection. It would be fair to say that Arab mathematicians believed that cubic equations as a whole could not be solved by algebraic means. This view passed to Europe with the mathematics transmitted by the Arabs. Individual cubics continued to be solved, however. Leonardo of Pisa considered a problem equivalent to the equation

$$x^3 + 2x^2 + 10x = 20$$

and obtained an approximate solution of surprising accuracy. There is an anonymous thirteenth-century Italian manuscript in which cubics of the form

$$ax^3 = cx + d$$

and

$$ax^3 = bx^2 + d$$

are considered, but the proposed methods of solution are incorrect. It is small wonder that in 1494 Pacioli in his *Summa* simply states that cubics cannot be solved algebraically. The *Summa* was a highly influential work and hence with this statement Pacioli was effectively throwing out a challenge which the sixteenth-century Italian mathematicians simply could not ignore.

We find a few examples of quartic and even sixth-degree equations in Babylonian mathematics, but these were of a special kind which could be treated as quadratics. These higher order equations arose as a result of Babylonian methods of tackling simultaneous equations. If we replace x by x^2 in a typical quadratic equation of the time, for example one of the type

$$x^2 + bx = c$$

we obtain the quartic equation

$$x^4 + bx^2 = c$$

which can be solved as a quadratic with x^2 instead of x as the unknown. Quartic equations of this kind could be solved by reference to tables in just the same way as quadratics. Having obtained the value of the unknown x^2, further reference to a table of square roots immediately yielded the value of x.

Quartic equations with their solutions appear in the works of Bhâskara and Mahâvîra, though once again only specific cases are considered and there is no attempt at discussing a general theory. A typical example occurs with the problem:

> What number which, multiplied by 200 and added to its square, then multiplied by 2 and subtracted from its fourth power, will become one myriad less unity?

This is just the equation

$$x^4 - 2x^2 - 400x = 9999$$

The solution described is ingenious. We have to add to both sides

$$4x^2 + 400x + 1$$

This gives us

$$x^4 + 2x^2 + 1 = 4x^2 + 400x + 10000$$

or

$$(x^2 + 1)^2 = (2x + 100)^2$$

We now take square roots, and obtain

$$x^2 + 1 = 2x + 100$$

that is,

$$x^2 - 2x - 99 = 0$$

and so $x = 11$ from the appropriate formula. It is small wonder that we are then told that 'in similar cases, the value of the unknown must be found by the ingenuity of the mathematician'. A systematic method of solution was discussed for the special case where all the coefficients of the unknown and its powers were equal, that is, problems leading to equations of the type:

$$a(x^n + x^{n-1} + \ldots + x^2 + x + 1) = k$$

The expression in brackets is a geometric progression with the common ratio x. The rule for solving such an equation is stated in the form:

> That by which the sum divided by the first term is repeatedly divisible, subtracting unity every time, is the common ratio.

One example yields the quintic equation

$$x^5 + x^4 + x^3 + x^2 + x + 1 = 1365$$

We now make a guess, $x = 4$, and calculate successively

$$(1365 - 1) \div 4 = 341$$
$$(341 - 1) \div 4 = 85$$
$$(85 - 1) \div 4 = 21$$
$$(21 - 1) \div 4 = 5$$
$$(5 - 1) \div 4 = 1$$
$$(1 - 1) \div 4 = 0$$

The solution is therefore 4. This method is, of course, hardly general since it is in part a trial-and-error method, involving making an appropriate guess for the value of x before the method is put into operation.

Quartic equations appear in Arab works, but the methods of solution, as with cubics, are geometric. The situation at the start of the sixteenth century was thus very much the same as with cubics. Apart from a few special cases, no purely algebraic methods of solution were known and by-and-large such solutions were thought not to be possible.

We now encounter a number of colourful figures who played out an Italian mathematical drama of a kind that was only possible in the Renaissance period. The chief actor is Gerolamo Cardano – mathematician, physician, philosopher, astrologer, and inveterate gambler – author of the *Ars magna* and a number of other works, including the first systematic mathematical treatment of games of chance. The *Ars magna* describes methods for the solution of cubic equations, and established Cardano as a leading figure in the mathematical world of the sixteenth century. However, the methods were not his own and it would appear that they had been well-kept secrets of others until Cardano was able, either by threats or bribery, to extract them from those who knew them. The story is somewhat tortuous and we cannot be certain of all its details.

It seems that sometime around 1515, Scipione dal Ferro, Professor of Mathematics at Bologna, discovered an algebraic method for solving cubic equations of the type

$$x^3 + cx = d$$

and revealed it only to one of his pupils, a certain Antonio Fior. Secrecy in such matters had to be the order of the day because university appointments did not normally have tenure, and renewal of a contract often depended on winning competitions in solving mathematical problems. The method of solution involved substituting $u - v$ for x. (We must remember, of course, that we are using modern notation here. In the sixteenth century such an equation would still be stated in words as 'the cube and things equal to a number'. It is only if we keep this in mind that we can appreciate the real difficulties which were involved in solving equations.)

To follow dal Ferro's method, we will work through an example.

Suppose that we have to solve the equation

$$x^3 + 9x = 6$$

If we substitute $u - v$ for x according to dal Ferro's rule, we have

$$(u - v)^3 + 9(u - v) = 6$$

that is,

$$u^3 - v^3 - 3uv(u - v) + 9(u - v) = 6$$

Now, if $uv = 3$, we can get rid of all the terms except for

$$u^3 - v^3 = 6$$

and we also have

$$u^3v^3 = 27$$

This is a very familiar situation when we regard u^3 and v^3 as new unknowns p and q, for we have

$$\left.\begin{aligned} p - q &= 6 \\ pq &= 27 \end{aligned}\right\}$$

In this case, we can see by inspection that $p = 9$ and $q = 3$ satisfy these conditions. We can, however, always solve such a pair of equations because substituting from one into another yields a quadratic equation and we can solve any quadratic just by using the formula. Dal Ferro could, of course, solve quadratics – his efforts were directed to solving cubics. So we have

$$\left.\begin{aligned} p = u^3 &= 9 \\ q = v^3 &= 3 \end{aligned}\right\}$$

giving us

$$\left.\begin{aligned} u &= \sqrt[3]{9} \\ v &= \sqrt[3]{3} \end{aligned}\right\}$$

and hence

$$x = u - v = \sqrt[3]{9} - \sqrt[3]{3}$$

Around 1530 another mathematician Nicola of Brescia known as Tartaglia (the Stammerer), a teacher in Venice, claimed to be able to solve cubic equations also. Around this time or perhaps a little earlier, he had discovered a method for solving cubic equations of the type

$$x^3 + bx^2 = d$$

By this time dal Ferro was dead, and Fior was able to exploit his secret method. He challenged Tartaglia to a public contest. Each was to provide 30 equations to be solved, and the loser would pay for a banquet to which the winner could invite his friends. The contest took place in 1535, and shortly before it Tartaglia also discovered for himself the method of solving cubic equations in which the term in x^2 is missing. Since Fior could solve only this latter type, Tartaglia won the contest.

Cardano heard of Tartaglia's victory and determined to persuade him to reveal his methods. He had been working on the problem of equations and had in fact made some progress, but his methods kept bringing him up against examples which required him to find square roots of negative quantities. This, of course, he was unable to do. After several unsuccessful attempts to worm secrets from him, Cardano invited Tartaglia to his house in Milan.

As well as being a teacher of mathematics, Tartaglia was also an engineer and something of an inventor. He was particularly interested in military science, and had written a book on artillery. Cardano undertook to introduce him to the Spanish Governor of Milan and to recommend his inventions if, in return, he would reveal his secret methods for solving cubics. The bargain was struck, but Cardano was required to swear an oath not to publish the methods, which Tartaglia had set out in some lines of verse which he gave to Cardano, beginning:

Quando che'l cubo con le cose appresso
Se agguaglia a qualche numero discreto:
Trovati dui attre differenti in esso.

(When the cube and the things add up to some whole number, take two others of a different kind.) It goes on to explain how to solve cubic equations by the substitution method already explained, and also by using a similar method, where x is put equal to $u + v$.

By 1545 Cardano and a pupil of his, Ludovico Ferrari, had amassed a large amount of material on both cubics and quartics which they wished to publish. There was, however, the matter of the oath. However, on looking through the papers which dal Ferro had left behind him at Bologna, they came across his solution of cubic equations of the type

$$x^3 + cx = d$$

On discovering that it was dal Ferro and not Tartaglia who had first discovered this, Cardano decided that he was justified in breaking his oath. He therefore published all his work in algebra, including Tartaglia's methods, in the *Ars magna*. The result of this was a series of acrimonious pamphlets written alternatively by Tartaglia and Ferrari, in which Tartaglia accused Cardano and Ferrari accused Tartaglia of plagiarism. So unpleasant was the whole atmosphere generated by this dispute that Tartaglia was lucky not to have been murdered by Cardano's supporters.

We have dwelt somewhat on this particular story, not merely because it is a good tale in its own right, but also because it throws light on the state of algebra in the sixteenth century and on the general social circumstances within which mathematical advances had to be made. In particular, the frequent reluctance of mathematicians to publish their discoveries had an inhibiting effect on these advances. Cardano was, of course, an exception to this.

The *Ars magna* also included material on quartic equations. Methods for solving these had been successful in particular cases, as we have seen, but little headway had been made in any general approach that was algebraic. About five years before the publication of the *Ars magna* another Italian mathematician, Zuanne de Tonini da Coi, had sent a problem to Cardano which leads to a quartic equation:

> Divide 10 into three parts such that they shall be in continued proportion and that the product of the first two shall be 6.

If we take the three parts to be w, x and y, we have

$$w + x + y = 10$$

$$w:x = x:y$$

whence

$$wy = x^2$$

and

$$wx = 6$$

If we eliminate w and y by substituting for them in terms of x, we obtain the equation

$$x^4 + 6x^2 + 36 = 60x$$

Cardano was unable to solve this equation, but Ferrari by solving it discovered a general method applicable to all quartic equations. This involved reducing the equation by an appropriate transformation to a cubic and, once this was done, the Tartaglia methods completed the solution.

A few years before Cardano died, yet another Italian mathematician, Rafael Bombelli, pointed out that in certain cases, even though a cubic equation in fact had three real roots, the methods of the *Ars magna* still led to expressions involving roots of negative numbers. Consider the equation

$$x^3 = 15x + 4$$

If we now substitute $u + v$ for x, we have

$$(u + v)^3 = (u + v) + 4$$

that is,

$$u^3 + v^3 + 3uv(u + v) - 15(u + v) = 4$$

This leads to

$$\left.\begin{aligned} u^3 + v^3 &= 4 \\ u^3v^3 &= 125 \end{aligned}\right\}$$

by the Tartaglia method. Now

$$\left.\begin{aligned} p + q &= 4 \\ pq &= 125 \end{aligned}\right\}$$

leads to the quadratic equation

$$p^2 - 4p + 125 = 0$$

which, by formula (or by completing the square) has the solution

$$p = 2 + \sqrt{-121}$$

whence

$$q = 2 - \sqrt{-121}$$

Hence the solution of the original cubic equation is

$$x = \sqrt[3]{2 + \sqrt{-121}} + \sqrt[3]{2 - \sqrt{-121}}$$

But we can see, virtually by inspection, that $x = 4$ is a solution. This is not immediately apparent from the Tartaglia result. Bombelli was able to show that it was possible to obtain a real root from awkward expressions such as that involving $\sqrt{-121}$ above, and thus effectively solved what up to then had been considered an irreducible case.

Towards the end of the century, Viète showed that the quadratic term can always be removed from a general cubic equation. This is achieved by substituting $u - (c/3)$ for x, where c is the coefficient of the square of the unknown. Reducing cubics in this way had been achieved long before in India, but it had not been proved that this could always be done. By the end of the seventeenth century, a little over a century after Viète's death, it was known that all quadratic, cubic and quartic equations have real or complex roots which can be found simply by substitution of the coefficients in appropriate formulae. It was also known that these equations had two, three and four roots respectively and that complex roots, where they existed, always occurred in pairs. The stage was well set for a concerted attack on higher degree equations.

By analogy, it was inferred that fifth, sixth, seventh, and higher degree equations must have five, six, seven, etc., roots. Could formulae be found for finding these? The number of roots is, of course, related to the number of ways in which an expression of the form

$$a_nx^n + a_{n-1}x^{n-1} + \dots + a_2x^2 + a_1x + a_0$$

(known as a polynomial) can be factorized into linear and quadratic factors, and we have already seen that methods of factorization go back at least to Thomas Harriot (early seventeenth century). Albert Girard, a contemporary of Harriot, was the first to have made the inference that every polynomial equation has at least one root, though this was later called into question by as great a mathematician as Leibniz. This theorem, eventually proved by Carl Friedrich Gauss in his doctoral thesis of 1799, is known as the 'fundamental theorem of algebra'. It had already been assumed by numerous mathematicians, including D'Alembert, Euler and Lagrange, each of whom had put forward 'proofs', subsequently shown to be incorrect. It can be shown to be equivalent to the statement that a polynomial of degree n has n roots. Gauss' thesis thus concluded one aspect of a long history of equations going back to Babylonian mathematics.

Searches for formulae by which roots of fifth and higher-degree equations could be found continued into the eighteenth century. Indeed, most eighteenth-century mathematicians, with a few notable exceptions such as Lagrange, expected that these formulae would soon be discovered. Eventually in the early part of the nineteenth century at Italian doctor put forward a proof that such formulae were not possible and this was confirmed for fifth-degree equations in 1824 by the famous Norwegian mathematician, Niels Abel, and for higher degree equations by Evariste Galois in 1831.

More than One Unknown

As with equations in only one unknown, those in more than one unknown which have to be solved simultaneously go back to the time of Ancient Egypt and Babylon, though surviving examples from Egypt are comparatively few, being confined to the Berlin and Kahum papyri.

We often encounter Babylonian examples where the sum and the product of two numbers are given. For example:

My length and breadth are together 6;30. My area is 7;30.

We are required to find the length and the breadth. If we write these as x and y, we have

$$\left.\begin{aligned} x + y &= 6;\,30 \\ xy &= 7;\,30 \end{aligned}\right\}$$

The solution described on the tablet requires us first to square half of $x + y$, that is, to square 3;15. From tables, this is found to be 10;33,45. We are now to subtract 7;30 from 10;33,45. This gives 3;3,45, the square root of which is 1;45. This is now to be added to and then subtracted from 3;15, giving us 5 and 1;30 which are, respectively, the required length and breadth. This method is equivalent to application of the formulae

$$x = \frac{x+y}{2} + \sqrt{\left(\frac{x+y}{2}\right)^2 - xy}$$

$$y = \frac{x+y}{2} - \sqrt{\left(\frac{x+y}{2}\right)^2 - xy}$$

The example involves solving a linear and a quadratic equation.

Babylonian systems of linear equations in two unknowns were usually solved by the method of substitution. However, another method, later to be known as the 'plus and minus' method was used when one of the equations involved just the sum of the two unknowns. For example, consider the problem:

> Half of my breadth taken from two-thirds of my length is 8,20. My length and breadth are together 30,0.

Half of $x + y$ is 15,0. So we take x to be $15{,}0 + u$ and y to be $15{,}0 - u$, and substitute into the first equation

$$\tfrac{2}{3}x - \tfrac{1}{2}y = 8{,}20$$

This gives us

$$1;10u = 5{,}50$$

from which $u = 5{,}0$. We now immediately obtain $x = 20{,}0$ and $y = 10{,}0$. Systems of linear equations with up to five unknowns are found in the tablets, and there is one astronomical problem involving as many as ten unknowns. We also find examples of simultaneous quadratic equations.

There is the problem:

> I have multiplied my length and my breadth and the result is 10. I have multiplied my length by itself. I have multiplied the excess of my length over my breadth by itself and the result by 9. This area is the area obtained from the length itself.

This gives us the two equations

$$\left.\begin{aligned} xy &= 10 \\ 9(x-y)^2 &= x^2 \end{aligned}\right\}$$

Substituting for y in the second equation we get the fourth-degree equation

$$8x^4 - 3{,}0x^2 + 15{,}0 = 0$$

This is a quadratic equation in x^2 for which the Babylonians had methods of solution. The roots are both positive, $x^2 = 15$ and $x^2 = 7;30$. The approximate square roots of these numbers were available in tables, and approximate values of y could then be found from $xy = 10$, again from tables.

Another example is the problem:

From one third of my length and my breadth I have subtracted one sixtieth of the square of the excess of my length over my breadth. The result is 15. My area is 10,0.

We have, therefore, the two equations:

$$\left.\begin{aligned} ;20(x + y) - ;1(x - y)^2 &= 15 \\ xy &= 10,0 \end{aligned}\right\}$$

No explanation of how the answer was obtained is given. However, if we put $x = u + v$ and $y = u - v$, we obtain

$$\left.\begin{aligned} 2u + 2v^2 &= 45 \\ u^2 - v^2 &= 10,0 \end{aligned}\right\}$$

Eliminating v^2 gives us

$$2u^2 + 2u - 20,45 = 0$$

and again we have a quadratic for which methods of solution were available. Finally, the appropriate multiplication table yielded v, and x and y followed.

Systems of first-degree and second-degree equations appear in Greek works also, sometimes involving more than two unknowns. Methods of geometrical solution were well-known; algebraic methods were much more rare. Three quadratic problems involving two unknowns appear in Euclid's *Data*. Thymaridas (fourth century B.C.) is said to have had this rule for solving a particular set of n linear equations in n unknowns:

If the sum of n quantities be given, and also the sum of every pair containing a particular quantity, then this particular quantity is equal to $1/(n + 2)$ of the difference between the sums of these pairs and the first given sum.

The rule, known as the 'flower of Thymaridas' thus gives the formula

$$x = \frac{(k_1 + k_2 + \ldots + k_{n-1}) - s}{n - 2}$$

for the solution of the set of equations

$$\left.\begin{aligned} x + x_1 + x_2 + \ldots + x_{n-1} &= s \\ x + x_1 &= k_1 \\ x + x_2 &= k_2 \\ \ldots \quad &\ldots \\ x + x_{n-1} &= k_{n-1} \end{aligned}\right\}$$

It is quoted in a work by Iamblicus (fourth century A.D.), who also describes how certain other systems of equations can be reduced to the form for which the rule is applicable.

Simultaneous linear and quadratic equations are found in Diophantus. He called the unknowns 'first number', 'second number', and so on, and this must have made the calculation of solutions highly confusing with almost everything written out in words, though he did introduce a minimum of algebraic notation.

Simultaneous equations receive considerable treatment in Indian documents. The special linear case in two unknowns, which we would write in the form

$$\left.\begin{aligned} ax + by &= c \\ bx + ay &= d \end{aligned}\right\}$$

is discussed by Mahavira in terms of calculating the price of fruit:

> From the larger price multiplied by larger number of things subtract the smaller price multiplied by the smaller number of things. Dividing by the difference of the squares of the numbers of things gives the price of each of the larger number of things. The price of the other will be obtained by reversing the multipliers.

In modern notation, the solution is given as

$$x = \frac{ac - bd}{a^2 - b^2}$$

$$y = \frac{ad - bc}{a^2 - b^2}$$

Equations of the form

$$x - \frac{c}{d}y = y - \frac{a}{b}x = k$$

are said to have the solution

$$x = \frac{b(c + d)}{b(c + d) - c(a + b)}k$$

$$y = \frac{d(a + b)}{d(a + b) - a(c + d)}k$$

In the Bakhshâlî treatise we find a set of equations of the form

$$\left.\begin{aligned} x_1 + x_2 &= k_1 \\ x_2 + x_3 &= k_2 \\ \ldots \quad &\ldots \\ x_{n-1} + x_n &= k_{n-1} \\ x_n + x_1 &= k_n \end{aligned}\right\}$$

where n is odd, solved by the rule of false. One of the problems solved in this way leads to the equations

$$\left.\begin{array}{l} x_1 + x_2 = 13 \\ x_2 + x_3 = 14 \\ x_3 + x_1 = 15 \end{array}\right\}$$

We are told to assume that $x_1 = 5$. Working through the three equations successively, this gives us $x_2 = 8$ and $x_3 = 6$. However, instead of the correct sum 15 in the third equation, we obtain 11. This has now to be adjusted according to the formula

$$x_1 = g + \tfrac{1}{2}(k_3 - k_3')$$

where g is the assumed value of x_1 and k_3' is the calculated value of k_3 based on this assumption. So, we have

$$\begin{aligned} x_1 &= 5 + \tfrac{1}{2}(15 - 11) \\ &= 7 \end{aligned}$$

and x_2 and x_3 follow immediately.

Although many of the Hindu methods of solution for simultaneous equations are general, that is, they are not explained solely in terms of particular equations, they relate only to special types of equations. There is no overall general method which will solve any set of n linear equations in n unknowns. The same is true of methods for solving problems leading to simultaneous quadratic equations. Special types are discussed by Âryabhaṭa, Brahmagupta, Mahâvîra, and others. Certain of these types were regarded as being particularly important, presumably because of the large number of everyday situations which led to such problems. One interesting case of this involves problems leading to the difference of the squares of two unknowns where the rules for solution are given the name 'dissimilar operations'. If we have at equations of the form

$$\left.\begin{array}{r} x^2 - y^2 = m \\ x - y = n \end{array}\right\}$$

the rule of dissimilar operations gives the solution as

$$x = \frac{1}{2}\left(\frac{m}{n} + n\right)$$

$$y = \frac{1}{2}\left(\frac{m}{n} - n\right)$$

The operations are called 'dissimilar' simply because in one formula we add n and in the other we subtract n. If the second equation has $x + y$ instead of $x - y$, we just reverse the terms inside the bracket in the formula for finding y. All these various solutions are, of course, explained in words.

Simultaneous linear equations had been solved in China from the earliest times from which we have mathematical records. Special methods of solution using bamboo rods were devised as early as 1000 B.C. In the *K'iu-ch'ang suan-shu* we find a number of problems leading to the solution of equations of the type

$$y = ax - b = cx + d$$

and the rule given for solution involves manipulation of bamboo rods representing the coefficients so as to obtain

$$x = \frac{b + d}{a - c}$$

$$y = \frac{ad + cb}{a - c}$$

This method required that the rods first be set out in an array:

$$\begin{matrix} a & c \\ b & d \end{matrix}$$

Thus we have a precursor of the determinant of quite extraordinary antiquity.

Our modern methods of solving sets of simultaneous linear equations using matrices and determinants date in Europe from the eighteenth-century works of Colin Maclaurin and Gabriel Cramer, though shortly before the end of the previous century the idea of setting out such equations in row and column form had been mentioned by Leibniz in a letter to the Marquis de l'Hôpital. Initially, only determinants were used; the modern theory of matrices dates from the work of James Sylvester in the middle of the nineteenth century, though logically the concept of a matrix comes first. We shall not discuss matrices in any detail. They are part of a branch of mathematics now generally known as 'linear algebra', beyond the scope of this book. Their importance for us is that with them a new kind of algebraic entity evolved. This no longer stood for a single number but for an array of numbers, and in working out how to manipulate such arrays, mathematicians were forced to accept a new and revolutionary concept – non-commutative multiplication. In the algebra of matrices, $A \times B$ is not generally the same as $B \times A$.

Non-commutative multiplication also arose in the manipulation of strings of numbers, known as n-tuples. These were effectively an extension of complex numbers, for the complex number $a + \sqrt{-1}\,b$ (usually written $a + ib$) can be regarded as a pair of numbers (a, b). Complex numbers arise, as we saw in the previous two sections, in the solution of equations and are manipulated just as ordinary numbers. The Irish mathematician, Willian Rowan Hamilton, investigated the extension of such number-pairs to triples (a, b, c) and quadruples (a, b, c, d), and presented a paper on the manipulation of quadruples (which he called 'quaternions') in 1843. This was the first non-commutative algebra to be developed.

Determinants arose first of all simply as arrays of the coefficients of simultaneous linear equations following the principle similar to that originally suggested by the Chinese mathematicians more than two-and-a-half millennia earlier. If we have two equations

$$\left.\begin{aligned} ax + by &= m \\ cx + dy &= n \end{aligned}\right\}$$

the determinant of the coefficients is written

$$\begin{vmatrix} a & b \\ c & d \end{vmatrix}$$

This is a two-by-two determinant – it has two rows and two columns. It is more convenient to use suffices which will tell us just where each coefficient is placed in the determinant. So we will re-write the equations as

$$\left.\begin{aligned} a_{11}x + a_{12}y &= n_1 \\ a_{21}x + a_{22}y &= n_2 \end{aligned}\right\}$$

and the determinant as

$$\begin{vmatrix} a_{11} & a_{12} \\ a_{21} & a_{22} \end{vmatrix}$$

We easily see now that the coefficient a_{21} is in the second row and first column. If we eliminate first y and then x from these equations we have

$$\left.\begin{aligned} a_{11}a_{22}x + a_{12}a_{22}y &= a_{22}n_1 \\ a_{12}a_{21}x + a_{12}a_{22}y &= a_{12}n_2 \end{aligned}\right\}$$

and, on subtracting,

$$(a_{11}a_{22} - a_{12}a_{21})x = a_{22}n_1 - a_{12}n_2$$

whence

$$x = \frac{a_{22}n_1 - a_{12}n_2}{a_{11}a_{22} - a_{12}a_{21}}$$

and a similar process gives us

$$y = \frac{a_{11}n_2 - a_{21}n_1}{a_{11}a_{22} - a_{12}a_{21}}$$

If we look closely at the formulae obtained for x and y, we see that both have the same denominator. This denominator can be obtained from the determinant of coefficients by multiplying diagonally as shown.

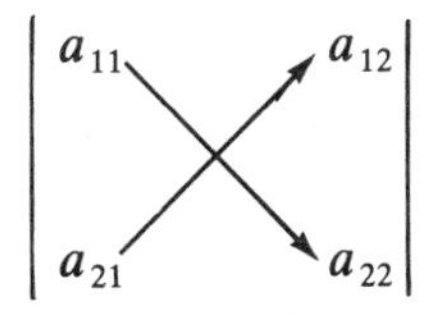

The positive term comes from the arrow pointing downwards and the negative term from the arrow pointing upwards.

If we now look at the numerators in the formulae, we find that the numerator for x is found by replacing the coefficients of x in the determinant by the numbers n_1 and n_2 on the right sides of the equations. Similarly, the numerator for y is obtained by replacing the y- coefficients by n_1 and n_2. Thus

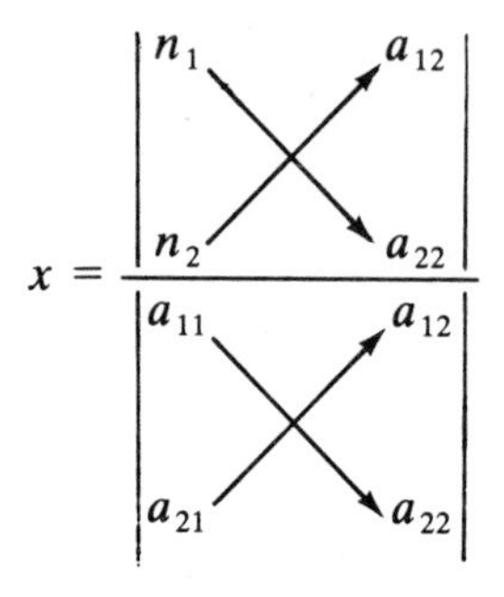

and

$$y = \frac{\begin{vmatrix} a_{11} & n_1 \\ a_{21} & n_2 \end{vmatrix}}{\begin{vmatrix} a_{11} & a_{12} \\ a_{21} & a_{22} \end{vmatrix}}$$

To see how this works out in practice, we will apply it to a specific example. Given the equations:

$$\left.\begin{aligned} 2x - 3y &= 1 \\ x + 2y &= 11 \end{aligned}\right\}$$

the determinant of the coefficients is

$$\begin{vmatrix} 2 & -3 \\ 1 & 2 \end{vmatrix}$$

We then have

$$x = \frac{\begin{vmatrix} 1 & -3 \\ 11 & 2 \end{vmatrix}}{\begin{vmatrix} 2 & -3 \\ 1 & 2 \end{vmatrix}} = \frac{2 - (-33)}{4 - (-3)} = \frac{35}{7} = 5$$

and

$$y = \frac{\begin{vmatrix} 2 & 1 \\ 1 & 11 \end{vmatrix}}{\begin{vmatrix} 2 & -3 \\ 1 & 2 \end{vmatrix}} = \frac{22 - 1}{7} = \frac{21}{7} = 3$$

This method of solving simultaneous linear equations can be extended to any number of equations. For three equations in three unknowns:

$$\left.\begin{aligned} a_{11}x + a_{12}y + a_{13}z &= n_1 \\ a_{21}x + a_{22}y + a_{23}z &= n_2 \\ a_{31}x + a_{32}y + a_{33}z &= n_3 \end{aligned}\right\}$$

the determinant of coefficients is

$$\begin{vmatrix} a_{11} & a_{12} & a_{13} \\ a_{21} & a_{22} & a_{23} \\ a_{31} & a_{32} & a_{33} \end{vmatrix}$$

To evaluate this according to the arrow method, it is convenient to write out the first two columns again on the right. Arrows indicating positive and negative products of the terms can then be written in:

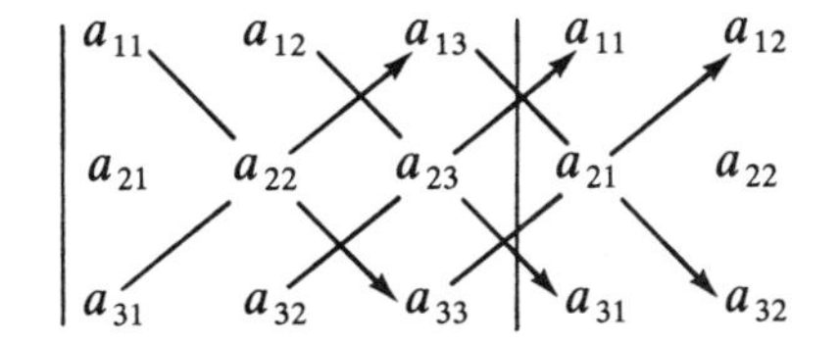

Three coefficients are multiplied together in each case. To solve for x, y and z we just replace the x, y, and z columns by the column of n's and evaluate the three determinants obtained, each being divided by the determinant of the coefficients.

There are two points of importance that we should note here. The first is that although we write out a determinant in the form of an array, it has a numerical value – that is, it can be regarded as just a number. The second is that the set of equations must be consistent and independent. Obviously, there is no solution to the set of equations

$$\left.\begin{aligned} x + y &= 3 \\ x + y &= 6 \end{aligned}\right\}$$

The sum of the two unknowns cannot be both 3 and 6 simultaneously. Again, we cannot calculate values for x and y if we can obtain one or more of the equations from the others as is the case, for example, with

$$\left.\begin{aligned} x + y &= 3 \\ 2x + 2y &= 6 \end{aligned}\right\}$$

where the second equation is obtained from the first by multiplying by 2. Any two numbers x and y, which together make 3 will satisfy both equations. Such equations are therefore indeterminate. Quite often, in practical problems, we encounter indeterminate equations where the circumstances of the problem from which they arise require that the solutions be whole numbers. Such equations are often called 'Diophantine'.

Indeterminate Unknowns

If we are presented with an equation of the form

$$ax + by = c$$

it is easy to see that we have been given insufficient information for us to find a unique solution. Thus, any two numbers adding up to unity satisfy the equation

$$x + y = 1$$

and there is an infinity of such pairs of numbers. We can rearrange it in the form

$$y = 1 - x$$

and, plotting values of y against corresponding values of x, we just obtain a straight line. Any point on this line represents values of x and y which satisfy the equation. The straight line is the locus of all such points. In this example x can be any real number, and there will always be a corresponding value of y.

Suppose that we have

$$x^2 + y^2 = 1$$

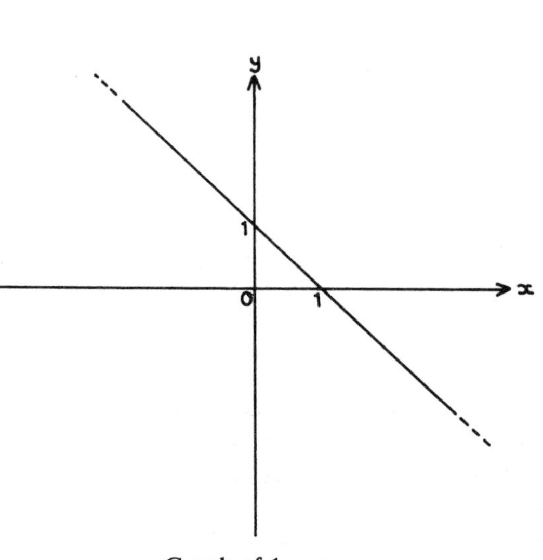

Graph of $1 - x$

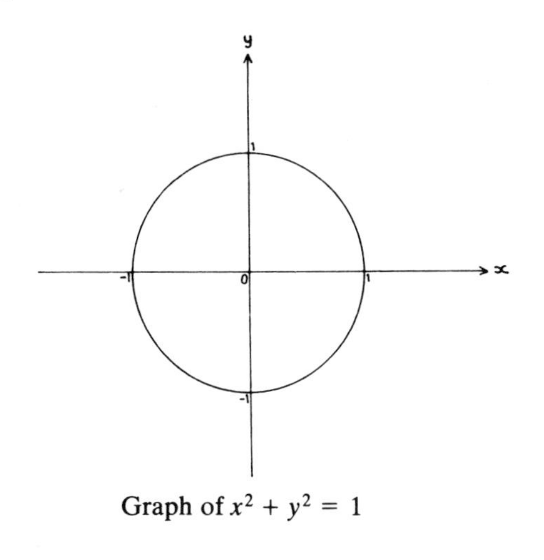

Graph of $x^2 + y^2 = 1$

Again, there is insufficient information for us to find a unique solution. Any two numbers whose squares add up to unity satisfy this equation. However, provided that we do not allow complex numbers to be considered, the values which x and y can take are restricted, though there is still an infinity of them. Plotting values of y against corresponding values of x yields a locus which is a circle of radius 1 centered at the intersection of the axis, that is, at the origin. This means that both x and y are confined to numbers from -1 to 1. For every value of x in this range, there will be two values of y, though these two values will coincide when $x = -1$ and $x = +1$.

Equations, such as

$$x + y = 1$$

and

$$x^2 + y^2 = 1$$

are examples of indeterminate equations. We simply do not have the information required to solve them, though we can find an infinity of individual values of x and y which satisfy them.

These are, however, very general forms of indeterminate problems. More often, we are presented with special restrictions arising from the circumstances from which a given problem has arisen. If we are discussing numbers of people we may well not want to have a solution in fractions – how do we interpret $1\frac{1}{2}$ persons? Many practical problems require us to find solutions which are positive whole numbers. Other problems may allow both positive and negative whole numbers, or may restrict us only to ratios of whole numbers. Yet again, it may be that we shall want solutions which are squares of whole numbers or fractions.

A considerable part of Diophantus' *Arithmetica* is devoted to indeterminate problems for which the solutions are required to be positive whole numbers or, at least, positive rational quantities. Both linear and quadratic problems of this kind are discussed and solved. More often than not whole number solutions are sought for the linear problems, fractional solutions being acceptable in the quadratic problems.

The oldest known problem of this kind arises with Pythagorean triples – three whole numbers x, y, z related in such a way that

$$x^2 + y^2 = z^2$$

This relation arises from the search for all possible right-angled triangles the lengths of whose sides are integral, the best known of which is the triangle whose sides are of lengths 3, 4 and 5 units. Various formulae were proposed by different Greek writers. According to Proclus, Pythagoras stated a rule for finding such triples which we would express as

$$n^2 + \left(\frac{n^2 - 1}{2}\right)^2 = \left(\frac{n^2 + 1}{2}\right)^2$$

where n is odd. An equivalent rule appears in the works of Plato.

Knowledge of Pythagorean triples goes back much earlier than the time of Pythagoras. A particularly famous Babylonian tablet, Plimpton 322, contains a list of such triples. It is the right-hand part of a larger tablet, the rest of which seems unfortunately to have been lost after it was excavated. Four columns are preserved and each has a heading – from right to left 'its name', 'number of the diagonal', 'number of the breadth', and an undeciphered phrase including the word 'diagonal'. Parts of the tablet have been worn away, and there is a piece missing on the right. A reconstruction of its con-

Babylonian tablet, Plimpton 322

d^2/h^2	b	d	n
1,59,0,15	1,59	2,49	1
*1,56,56,*58,14,50,6,15	56,7	3,12,1	2
*1,55,7,*41,15,33,45	1,16,41	1,50,49	3
*1,53,10,*29,32,52,16	3,31,49	5,9,1	4
1,48,54,1,40	1,5	1,37	5
1,47,6,41,40	5,19	8,1	6
1,43,11,56,28,26,40	38,11	59,1	7
1,41,33,59,3,45	13,19	20,49	8
1,38,33,36,36	9,1	12,49	9
1,35,10,2,28,27,24,26,40	1,22,41	2,16,1	10
1,33,45	45	1,15	11
1,29,21,54,2,15	27,59	48,49	12
1,27,0,3,45	7,12,1	4,49	13
1,25,48,51,35,6,40	29,31	53,49	14
1,23,13,46,40	56	53	15

Explanation of the tablet, Plimpton 322. Numerals in italics are missing from the tablet

tents has been possible, however, and it is agreed that the columns contain (from right to left) the numbers 1 to 15 (giving the row number), hypotenuses of right-angled triangles, bases of right-angled triangles, and squares of the corresponding heights. Thus, if we take d to denote a hypotenuse, b a base, and h a height, the headings of the respective columns are:

$$\frac{d^2}{h^2} \quad b \quad d \quad n$$

In any one row, the entries are related by:

$$d^2 - b^2 = h^2$$

We are thus faced with two interesting questions: 'what was in the missing columns?' and 'how were these numbers calculated?' We can only give speculative answers. There must have been a column of heights, and it is a straightforward operation to calculate what these heights were. Answering the second question is helped by the fact that the entries in the left-hand column (once they have all been reconstituted and errors in the other columns allowed for) start at 1,59,0,15 and decrease gradually to 1,23,13,46,40. Since the square of the height is the denominator of the heading, this means that we effectively have a succession of triangles arranged in order of increasing angle (between the base and the hypotenuse) from about 45° to about 60°. If we start with $h = 1$, we have

$$d^2 - b^2 = (d + b)(d - b) = 1$$

The numbers $d + b$ and $d - b$ are therefore reciprocal. If we now choose $d + b$ to be a regular sexagesimal number k (that is, one which has an exact sexagesimal reciprocal), we have

$$\left.\begin{aligned} d + b &= k \\ d - b &= \frac{1}{k} \end{aligned}\right\}$$

and hence:

$$d = \frac{k + \dfrac{1}{k}}{2}$$

$$b = \frac{k - \dfrac{1}{k}}{2}$$

A calculation of this sort provides the most plausible hypothesis as to how the numbers in the tablet were obtained. It is a typically Babylonian method. We cannot be certain that this hypothesis is correct, but the process of sorting and deciphering tablets is a continuing one and it may be that a tablet explaining the actual method used will be discovered.

Diophantus does not present general methods for solving indeterminate problems. Some of his ways of solving specific problems are, however, highly ingenious and are applicable to problems of similar types. He does not give the general solutions to specific equations either. He usually contents himself with presenting one solution. It is possible that this is because he was aware that in many cases, once one solution has been found, the others can easily be calculated from it.

Euclid's algorithm provides us with a method of solving linear indeterminate equations in two unknowns. To see how this works, we will consider the example

$$17x + 26y = 1$$

the requirement being that x and y be whole numbers – clearly, they cannot both be positive. Applying Euclid's algorithm, we have

$$\begin{aligned} 26 &= (1 \times 17) + 9 \\ 17 &= (1 \times 9) + 8 \\ 9 &= (1 \times 8) + 1 \\ 8 &= (8 \times 1) + 0 \end{aligned}$$

If we now begin to work backwards, we find first that

$$1 = (1 \times 9) - (1 \times 8)$$

then, on working back further and replacing 8 in this expression, we have

$$1 = (1 \times 9) - \{1 \times [(1 \times 17) - (1 \times 9)]\}$$
$$= (2 \times 9) - (1 \times 17)$$

on collecting together the multipliers of 9. We now replace 9 by working back again and obtain

$$1 = \{2 \times [(1 \times 26) - (1 \times 17)]\} - (1 \times 17)$$
$$= (2 \times 26) - (3 \times 17)$$

on collecting together the multipliers of 17. So $x = -3$ and $y = 2$ is an integer solution of the indeterminate equation.

However, we have found one solution only. There is more work to be done to find the complete solution. To see how we go about this, we return to a general form

$$ax + by = 1$$

in which we specify that a and b are relatively prime. If we have a solution $x = p, y = q$ then

$$ap + bq = 1$$

Subtracting the latter from the former, we have

$$a(x - p) + b(y - q) = 0$$

that is

$$a(x - p) = -b(y - q)$$

Now any factor of the left-hand side must also be a factor of the right-hand side. So a must be a factor of $(y - q)$ and b must be a factor of $(x - p)$, since a and b are relatively prime. Thus

$$x - p = bn$$

hence

$$y - q = -an$$

where n is any whole number. So the complete solution of the equation

$$ax + by = 1$$

$$\left.\begin{aligned} x &= p + bn \\ y &= q - an \end{aligned}\right\} n \text{ any integer.}$$

We can see that this is so by substitution in the original equation.

If we now return to the equation

$$17x + 26y = 1$$

and our solution $x = -3$, $y = 2$, we now have as the complete solution

$$\left.\begin{aligned} x &= -3 + 26n \\ y &= 2 - 17n \end{aligned}\right\} n \text{ any integer.}$$

There are thus infinitely many solutions.

Having considered the case where the right-hand side is unity, we are now in a position to consider the more general equation

$$ax + by = c$$

where we again seek whole number solutions and we assume that any factor, common to a, b and c has been divided out and that a and b are relatively prime. Let us suppose that we have used Euclid's algorithm to find a solution $x = p$, $y = q$ of the equation

$$ax + by = 1$$

so that

$$ap + bq = 1$$

and

$$\left.\begin{aligned} x &= p + bn \\ y &= q - an \end{aligned}\right\} n \text{ any integer}$$

is the general solution. If we multiply these general values of x and y by c, we will now obtain the general solution of

$$ax + by = c$$

However, as n is any integer, there is no difference between writing cbn and bn or between writing can and an. We can therefore state the general solution to be

$$\left.\begin{aligned} x &= cp + bn \\ y &= cq - an \end{aligned}\right\} n \text{ any integer.}$$

In our discussions, we have made a point of requiring that a and b be relatively prime. There is a reason for this. Once we have divided through by the common factors, if any, of a, b, and c and thus reduced the given equation to its lowest terms, the existence of any remaining common factor on the left-hand side would mean that the equation has no solution. If we have two equal numbers, one of which is $ax + by$ and the other c, they must have the same factors. So a common factor of a and b, which must in turn be a factor of $ax + by$, must also be a factor of c; yet we have excluded this possibility by requiring that the equation be in its lowest terms. If, therefore, we are asked to solve the equation

$$6x - 12y = 21$$

we first note that 3 is a common factor of 6, 12, and 21, so we divide through by 3 and obtain

$$2x - 4y = 7$$

We are now in a situation where 2 and 4 have a common factor, 2, which is not also a factor of 7. Clearly, the left-hand side must necessarily always be even, so the equation can have no whole number solution – the difference of two even numbers cannot be odd.

Often, practical problems which require that whole number solutions should be positive thereby limit the number of possible solutions. This must always be the case when a, b and c are positive, since eventually the left-hand side is bound to exceed c. As an example we solve the equation

$$8x + 20y = 364$$

for positive whole-number values only. First, we note that we can divide through by 4. This gives:

$$2x + 5y = 91$$

Whole-number solutions do exist, since 2 and 5 are relatively prime. We can see by inspection that $x = -2$, $y = 1$ is a solution of

$$2x + 5y = 1$$

hence the general solution of our original equation is

$$\left.\begin{aligned} x &= -182 + 5n \\ y &= 91 - 2n \end{aligned}\right\} n \text{ any integer.}$$

If x and y are both to be greater than zero, then

$$\left.\begin{aligned} -182 + 5n &> 0 \\ 91 - 2n &> 0 \end{aligned}\right\}$$

that is,

$$n > 36\tfrac{2}{5}$$

and also

$$42\tfrac{1}{2} > n$$

The whole number values of n which lie between $36\frac{2}{5}$ and $42\frac{1}{2}$ are 37, 38, 39, 40, 41, 42. Thus our equation has exactly six positive solutions for x and y, which we obtain by substituting these six values for n in our general solution.

We have discussed the solution of linear problems in two unknowns in some detail because, once the principles involved have been understood, it is not difficult to extend them to linear problems in more than two unknowns, though we shall not do this here. Such problems were clearly considered to be trivial by Diophantus, who was much more interested in finding squares and cubes of numbers which have special properties. Finding Pythagorean triples is just one example of this. Many of his problems eventually involve such triples during the course of their solutions, which were not usually restricted to whole numbers. The problems discussed in the *Arithmetica* were not all due to Diophantus himself. The work is clearly a collection; Diophantus must have drawn on a number of sources, though some of the problems are of his own invention.

Problems involving Pythagorean triples were also solved from very early times in India and China. In India, indeterminate analysis was considered so important that at one time the whole art of solving equations was named after it. We find the general solution of

$$ax + by = c$$

and of

$$ax - by = c$$

for positive whole numbers in the works of Âryabhata, Bhâskara, Brahmagupta and others. We also find quadratic indeterminate problems involving as many as four unknowns. Some of these problems are of greater difficulty than any of those to be found in Diophantus' *Arithmetica.*

Knowledge of Greek and Hindu indeterminate analysis passed to the Arabs, a number of whom included this topic in their own mathematical works. The most important of these was the *Fakhrî*, written in the eleventh century by Al-Karkhî. Whole portions of the *Arithmetica* are included in this work, but Al-Karkhî adds many problems of his own and extends a number of Diophantus' individual methods of solution to new types of problem. Some of this material found its way into the *Liber abaci* of Leonardo of Pisa.

Leonardo became something of an expert with indeterminate problems. In 1224, he was summoned by the Emperor Frederick the Second to take part in a mathematical contest, the problems for which were to be set by John of Palermo. These included indeterminate problems, all of which Leonardo was able to solve without difficulty. Subsequently in another work, the *Liber quadratorum*, he presented a number of new methods of solution which differ from those of Arab and Greek mathematicians, and represent a further advance in the development of this kind of analysis.

Although problems to be found in the *Arithmetica* became known in the West through Leonardo's writings and through Latin translations of Arab works, we do not find any specific reference to Diophantus himself until the fifteenth century. A number of editions of the *Arithmetica*, which had been discovered in the library at the Vatican, were published in the sixteenth and seventeenth century. A copy of the most famous of these, published by Bachet de Méziriac in 1621, came into the possession of the French mathematician Pierre de Fermat. Fermat used the margins of this work to jot down comments and results as they occurred to him. It is here that we find for the first time one of the great unsolved problems of number theory. Fermat had proved earlier that there are no positive integers x, y and z such that

$$x^3 + y^3 = z^3$$

In the margin of Bachet's *Diophantus*, he wrote that he had a proof of an extension of this, namely that there are no positive integers x, y and z such that

$$x^n + y^n = z^n$$

for any integral n greater than 2. Unfortunately he did not state this proof, contenting himself with adding only that it was one 'which this margin is too narrow to contain'. Neither his proof nor any other proof has ever been discovered, and Fermat's 'great theorem' (as it is called) remains a conjecture to this present day.

One of the most famous indeterminate equations studied by Fermat is

$$x^2 - by^2 = 1$$

This equation is often incorrectly attributed to John Pell. It was known to the Greeks and occurs in problems discussed by Theon of Smyrna and others of the late Greek period. One of the earliest problems leading to an equation of this type is the 'cattle problem', attributed by Arab writers to Archimedes, but almost certainly of a later date. It requires us to find the numbers of bulls and cows of four colours, white, blue, yellow and piebald, satisfying certain conditions. If we represent the required numbers as W, B, Y, P (bulls) and w, b, y, p (cows), the given conditions can be stated as

$$\left.\begin{aligned} B &= (\tfrac{1}{4} + \tfrac{1}{5})(Y + P) \\ P &= (\tfrac{1}{6} + \tfrac{1}{7})(W + Y) \\ w &= (\tfrac{1}{3} + \tfrac{1}{4})(B + b) \\ b &= (\tfrac{1}{4} + \tfrac{1}{5})(P + p) \\ p &= (\tfrac{1}{5} + \tfrac{1}{6})(Y + y) \\ y &= (\tfrac{1}{6} + \tfrac{1}{7})(W + w) \end{aligned}\right\}$$

Substitution eventually leads to the equation

$$x^2 - 4729494y^2 = 1$$

It has been claimed that to find the values for the eight unknowns in just one of the possible solutions would take more than six-hundred pages.

Examples of the so-called Pell equation are to be found in the *Arithmetica*, though there is no detailed discussion of them. It is possible that Diophantus considered such problems further in one of his works known to be lost. We also find specific equations solved in the works of Bhâskara. Fermat's main contribution was to show that when b is not a perfect square there are an infinite number of whole-number solutions. The Pell equation was later studied in the eighteenth century by Euler and subsequently by Lagrange, who published a completely general solution covering all possible cases.

Today, number theory is still a significant area of pure mathematics, though its importance is declining. The advent of computers has enabled many solutions to specific equations to be computed which would not have been found by other means. A number of interesting problems remain unsolved, however, of which the proof of Fermat's 'great theorem' is a notable example. Although the study of whole-number solutions of equations is now of less interest than it was a number of centuries ago, it is still important for those many practical problems for which fractional and irrational solutions are meaningless. This has, as we have seen, a very long history for much of which mathematicians lacked the convenient algebraic notation available today. For this reason alone, we should recognize just how great were the achievements of Diophantus and other mathematicians who long ago laid the foundations of modern number theory.

Symbols for the Unknown

We have already commented on a number of occasions about the lack of available algebraic symbolism in early times. We are so familiar with the idea that x designates an unknown quantity that we forget that this has been common only since the time of Descartes (seventeenth century). Writing equations and working out solutions was therefore for much of the time a long-winded affair, and cannot but have impeded the development of algebra.

We usually distinguish three stages in the evolution of algebraic notation and speak of written algebra as 'rhetorical', 'syncopated' or 'symbolic'. It is rhetorical when it is expressed entirely or almost entirely in words. The syncopated stage is reached when there are abbreviations and a limited amount of symbolism (usually resulting from abbreviations). Modern algebra is entirely symbolic in that equations are written down without any use of words.

Part of the symbolism employed in algebra is common to both algebra and arithmetic. We have already discussed the evolution of symbols for numbers and for the operations of arithmetic. Here, we shall just be concerned with symbolism associated with algebra, and especially with the representation of the unknown quantities.

Egyptian, Babylonian and Greek algebra before Diophantus was all largely rhetorical. We do find traces of what we would term 'syncopated' notation in the Egyptian mathematical papyri, but it is purely arithmetic except in the case of the hieratic symbol used to denote 'heap'.

Diophantus introduced the use of the terminal *sigma*, that is ς, to denote the unknown. We are not certain just why this particular letter was chosen. It is possible that it was because it was the only Greek letter not used as a numeral. Alternatively, it may have derived from the last letter of *arithmos*, a word often used in the sense of 'unknown number'. When the unknown was multiplied by a number, Diophantus used $\varsigma\varsigma^{οι}$ followed immediately by the letter denoting the multiplier, thus what we should write as $3x$ appears as

$$\varsigma\varsigma^{οι}\gamma$$

We have already noted his used of Δ^{Y}, K^{Y} and so on, for powers of the unknown.

Indian algebra was, like that of Diophantus, syncopated. There was a tendency to make use of the first syllables of words as algebraic abbreviations. Brahmagupta used *yâ* (from *yâvattâvat* meaning 'so much as') when there was only one unknown and then the first syllables of words for colours for further unknowns – *kâ* (from *kâlaka* meaning 'black') for example. Bhâskara also used this kind of abbreviation. Sometimes the word *sunya*, which also meant 'zero', was used for the unknown, and this led to the practice of using a dot (the early zero symbol) which we find in the Bakshâlî treatise. Sometimes abbreviations of the ordinal numbers were used for unknowns: *pra* (from *prathama*, 'first'), *dvi* (from dvitîyâ, 'second'), *tṛ* (from *tṛ*tîya, 'third'), and so on. We also find abbreviations of words denoting fruit, flavours and jewels.

It is surprising that, although the Arabs were quick to take over the Hindu numerals, they did not make use of such algebraic notation as had already been introduced in the *Arithmetica* and in the Indian writings. Of course, most of this notation came from Hindu and Greek words and would not have been meaningful in the same way to Arab readers. This may well be the explanation of why the Arab mathematicians returned to the rhetorical algebra of earlier times. The principle of using abbreviations of words was, however, plain for them to see, and their reasons for rejecting this principle remain unknown. It could easily have been adapted to their own language.

Al-Khwarizmî, Al-Karkhî, and other Arab mathematicians designated the unknown simply as 'thing'. This was translated directly into Latin as *res* and into Italian as *cosa*. It was from this that algebraists came to be known as 'cossists'. We find *res* in the work of Leonardo of Pisa and *cosa* in Pacioli's *Summa* and in the works of later Italian writers. When there were more than one unknown, these were known as 'first thing', 'second thing', and so on – a nomenclature to which Pacioli took objection. He himself uses two forms of notation, both abbreviations of words. He uses *co.* (from *cosa*) for the unknown, *ce.* (from *censo*) for the square of the unknown, *cu.* (from *cubo*) for its cube, and *cece.* (from *censocenso*) for its fourth power. However, he also uses ℞.p^{a}, ℞.2^{a}, ℞3^{a} (from *radix prima*, *radix secunda* and *radix terza*) and so on for our x^0, x, x^2, etc.

The beginning of the break-through came from Viète. In *In artem*, he made use of vowels to denote the unknowns and consonants to denote constants. Thus, what we write as

$$x^2 + 2bx = c$$

Viète wrote as

A quad. + B2 in A, aequatur Z plano.

His A quad. (for A *quadratum*) and A cubum were soon shortened to Aq. and Ac. This represents a significant departure of principle since the letters used now bear no relation to words; they are purely symbolic. In fact, Viète was not the first to adopt this symbolic principle, though he was the first to create a system from it by means of which equations could be written down in a general form. Stifel used a variety of algebraic notations including A for the unknown, and AA, AAA, and so on for powers of the unknown.

By the early part of the seventeenth century there were so many different algebraic notations in use that the Italian mathematician Pietro Cataldi attacked the confusion which was being created for writers, readers and printers alike and proposed that 0̸, 1̸, 2̸, 3̸, and so on should stand for what we would write as x^0, x, x^2, x^3, etc. This worked, however, only for equations in one unknown and consequently failed to appeal to other mathematicians. It is of passing interest because some historians have mistakenly seen in Cataldi's 1 the x which was to inspire Descartes' symbolism for algebra.

It was Descartes who, in his *La géométrie* (1637), eventually introduced the modern convention which we use today – letters from the end of the alphabet to denote unknowns; letters from the beginning of the alphabet to denote constants – though he was not always consistent in the use of x^2 for the square of the unknown x, sometimes writing xx. He was also frequently inconsistent writing to other mathematicians. Several years after the publication of *La géométrie*, we find him writing

$$1\mathrm{C} - 6\mathrm{N} = 40$$

rather than the

$$x^3 - 6x = 40$$

which would be in accord with his stated principles. We also find him using capital letters $A, B, C, \ldots$ for unknowns and small letters $a, b, c, \ldots$ for constants.

Descartes' modern notation was not immediately popular with all his contemporaries. We find *cossa* used as late as 1670, and largely syncopated algebras appeared well into the eighteenth century. Descartes notation was used in Gerard Kinckhuysen's *Algebra* (1661), in de la Hire's *Nouveaux elements des sections coniques* (1679), and by the turn of the century it had, with only a few exceptions, become the norm for mathematicians throughout western Europe. We may well wonder why it had not appeared much earlier than it did. The answer would seem to lie in the habit of expecting mathematical abbreviations to retain some obvious link with what they were signifying and especially with the spoken word. The older Greek works, which inspired so much of the mathematics of the Renaissance period and after, were purely rhetorical. Mathematical ideas were explained in words; mathematical arguments were written in words. To adopt abbreviations of words is therefore a natural step; the change to abstract symbolism demands an intellectual leap of extraordinary magnitude. It is precisely this requirement to write and hence to think in terms of symbols which makes mathematics a difficult subject in the classroom today unless attention is paid to this particular intellectual demand.

Chapter Seven
Recreational Numbers

Let schoolmasters puzzle their brain.
(Goldsmith)

Rationalism for the few and magic for the many.
(Burckhardt)

Numbers have long had a fascination of their own quite apart from their practical application. Many of the mathematical writers whose works have come down to us stress that their problems are for pleasure as well as for intellectual edification. Playing around with numbers has been a pastime for the past 3000 years and perhaps for much longer. Many modern conundrums and children's games have their origins in early problems of mathematics. Sometimes they appear dressed in radically differing guises; sometimes the actual numbers involved are changed around. Not infrequently we find very little change even though the problem concerned has travelled across continents and through centuries of history.

From time to time, collections of recreational problems have appeared. On other occasions, recreational problems have been included, along with those of a strictly practical nature, in compendia of elementary arithmetic and algebra. We find them in the *Greek Anthology*, in Leonardo of Pisa's *Liber abaci*, in Pacioli's *Summa*, in Chuquet's *Triparty*, and in many other works. Some nineteenth- and twentieth-century anthologies have been compiled from numerical 'teasers' in newspapers and magazines.

One recurring problem is that of the man with seven wives, each wife having seven sacks, and so on. Another is the problem of ferrying a wolf, a goat and a cabbage across a river. The boat will hold only one of these in addition to the man who has to do the ferrying. The restriction imposed is that the wolf cannot be left unattended with the goat, nor the goat with the cabbage. This is not strictly a numerical problem, though it has frequently appeared in collections of numerical problems. Various extensions have been proposed. In one version, there are three married couples wanting to cross the river, the restriction imposed being that no wife can be left without her husband if another man is present. Such problems can be solved with a little logical thought. Mathematically, they belong to a class of problems known as

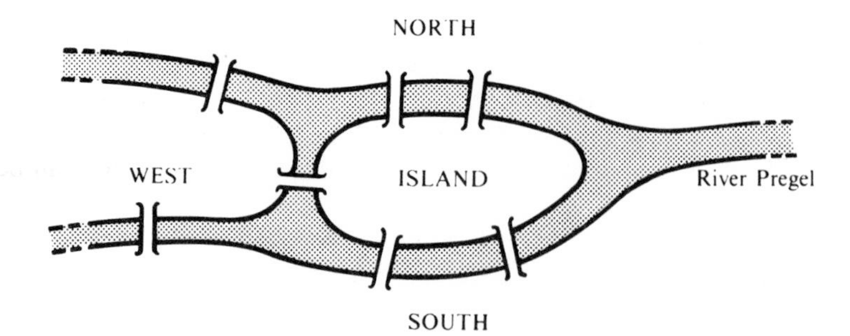

The Koenigsberg Bridge Problem: Is it possible to visit each of the four parts of the city crossing each bridge once and once only?

'combinatorial' and we can solve them by graph theory. One of the most famous graph-theory problems is that of the Koenigsberg bridges. This was first solved by Leonhard Euler in 1735, but it may date from the seventeenth century or even earlier, though not necessarily in Euler's precise form.

Some numerical problems are posed using letters. These are not necessarily anything to do with gematria. A typical example is:

Complete the addition:

```
 J F G B A
 E D I G H
 ---------
 - - - - E
```

Solutions to such problems are not always unique. Usually, they require logic rather than calculating skill. Here, we can immediately infer that E + J cannot be greater than 9. Indeed, if a carry from the previous place is involved, E + J cannot be greater than 8. We can infer also that E cannot be less than 3 and that, if E is equal to 3, A and H must be 1 and 2 (or 2 and 1). So we have already made a significant start with the solution.

Many simple games depend upon numerical reasoning. One of the best-known of these is the game of Nim. In its simplest form it is a game in which two players in turn remove up to a specified number of matches from a heap. The player forced to remove the last match is the loser. Clearly, the maximum number of matches which can be removed at one turn is crucial to determine what the strategy is. Suppose that the maximum which can be removed at a turn is 4. If we are faced with 2, 3, 4 or 5 in the heap, we can leave our opponent with just one to pick up. To be faced with 6 matches, is however, to be in a losing position, since however many we pick, we are bound to leave our opponent with 2 to 5 matches. We can now continue this argument: 7 to 10 is a winning situation since our opponent can be left with 6; 11 is a losing situation since we are forced to leave our opponent with 7 to 10, and so on. The losing situations are $5n + 1$, where n is zero or any positive whole number. Our aim at any turn must be to leave our opponent with $5n + 1$ matches.

There are all sorts of variations of the game of Nim. One popular version begins with three heaps of matches and any number can be removed provided that these are taken from one heap only. Clearly, being faced with 0, 0, 1 or 1, 1, 1 or 0, 2, 2, are all losing situations and hence situations with which one wants to face one's opponent. An argument for a winning strategy can begin here.

Many of the problems to be found in anthologies are indeterminate. In a collection attributed to Alcuin, we find the problem:

> 100 bushels are distributed among 100 persons so that each man receives 3 bushels, each woman 2 bushels, and each child half a bushel. How many men, women and children are there?

This has more than one solution, but, like Diophantus, the author gives only one of them. A preference for problems involving 100 as a total is to be found amongst many Renaissance compendia and this preference can be traced back to earlier Arab, Hindu and Greek works. Sometimes, the problems are specifically designed so as to appear as if insufficient information has been given. A fairly modern example, but dating from before the decimalization of the British coinage, is as follows:

> There are 100 seats in a theatre. Half the men in the audience paid 10s. each for their seat; two-thirds of the women paid 7s. 6d. each; three-quarters of the children paid 6s. 8d. each. The rest of the audience were on free passes. If the theatre was full, how much was taken at the box office?

Here again we have the type of problem whose solution depends on some logical reasoning before any calculation can be attempted. There is sufficient information given for a unique solution to be found, though at a first glance people are apt to claim that this is not so. Numerically, it is just a problem of inverse proportion. There is, however, the little matter of first seeing how the proportions quoted (half, two-thirds and three-quarters) are related to the corresponding amounts paid. Alternatively, we can just argue that the numbers of men, women and children present obviously do not matter, and so assume any numbers which are convenient for calculation – say, 100 men and no women or children.

Recreational number problems are part of the folklore of most areas of the world with the one exception of the Americas where they are found only amongst the peoples of European, African or Far-Eastern origin. The Africans and the Chinese in particular provide a rich and rewarding hunting-ground for seekers of numerical games and teasers. It is not possible to present even a representative survey of such games and puzzles here, to do so would take up a complete volume by itself, and there are a number of books already available which cover the ground reasonably well. We therefore choose a few special areas for discussion, the aim being just to present something of the flavour of recreational mathematics as it has existed for many centuries.

'Think of a Number'

Most of us will probably have been faced at one time or another with a mathematical conundrum beginning 'think of a number'. We are apparently given an entirely free choice of any whole number. We are then told to carry out various operations, at the end of which we are informed by our questioner exactly what the result of our calculations is.

We do not have to think very deeply to appreciate that in these conundrums we can be given an apparently free choice of number because the one which we choose is irrelevant to the final outcome of the calculations. In the course of these it is carefully arranged that our original number is removed from the scene so that we are eventually calculating only with numbers selected by our questioner. A very little algebra quickly reveals the trick. Indeed, 'think of a number' problems, although they appear to involve only arithmetic, are algebraic in principle. It is probable that algebra had its beginnings in just such problems.

A typical example goes as follows:

Think of a number.
Double it.
Add 10.
Treble the result.
Subtract 15.
Divide by 3.
Take away twice the number first thought of.
The answer is 5.

To see that this is not magic but the application of a little simple algebra, all that we have to do is to follow the same instructions through, taking n to be the number first thought of. We then successively have

$$n$$

$$2n$$

$$2n + 10$$

$$6n + 30$$

$$6n + 15$$

$$2n + 5$$

$$5$$

The principle is quite clear. As soon as we take away twice the number first thought of, we are entirely in the questioner's hands. He can go on to tell us to carry out all sorts of operations, but, once we have reached 5, our original choice of number has become irrelevant.

There are a number of variations on this theme. Sometimes we are asked to state the result of carrying out a series of prescribed operations on a secretly chosen number. From this stated result, our questioner is then able to deduce our original choice. Sometimes we are invited to choose more than one number. Occasionally we are asked to look up numbers in a telephone directory or to choose page numbers, dates of birthdays, or other numbers supposedly known only to ourselves.

An example using more than one chosen number is the following:

> Take the last three digits of your telephone number.
> Multiply this number by 10.
> Add 43.
> Add the last digit of your house number.
> Multiply by 10.
> Add the number of brothers and sisters which you have.
> Take away 430.

The three numbers chosen then form the number resulting from the calculation, for if the numbers are a, b and c, the answer is $100a + 10b + c$. The problem does assume, however, that we were not born into a family of more than 10. Problems of this kind are very easy to devise and can provide amusement for adults and children alike.

Two of the problems to be found in the Rhind papyrus are of the 'think of a number' type. Problem 28 reads:

> Two-thirds is to be added.
> One-third is to be subtracted.
> 10 remains.
> Take one-tenth of this; there becomes 1.
> The remainder is 9.
> Two-thirds of it, namely 6, is to be added.
> The sum is 15.
> One-third of this is 5.
> Lo! 5 goes out; remainder 10.
> This is how it occurs.

A possible interpretation is:

> Think of a number.
> Add two-thirds of it.
> Take away one-third of this sum.
> Let 10 remain.
> One-tenth of this is 1.
> 1 from 10 leaves 9.

The question asks for the result of the calculation after the first two operations, and from this result, 10, deduces that the original number chosen was 9. The rest of the problem provides a 'proof' – that is, a numerical verification. Clearly, this works for any number n which we may care to think of. We add ,n and then take away one-third of the result. This leaves us with $\frac{10}{9}n$. Taking away one-tenth of this will always give us n. Problem 29 of the papyrus is similar and is written alongside Problem 28. They may be the earliest 'think of a number' problems known to us, though by no means all of the scholars agree with this particular interpretation of them. Problems which can be interpreted in this way can also be found in Diophantus and in some of the Hindu mathematical writings, but it is in early European Renaissance works that they begin to be stated explicitly, for example, in the *Tractatus de numeris datis* of Jordanus Nemorarius.

Tartaglia discussed 'think of a number' problems in the *General trattato de numeri e misuri*, published in two parts in 1556 and 1560. They also appear in Claude Bachet's *Problèmes plaisants* of 1612, an enlarged edition of which appeared in 1624. Jacques Ozanam considered them in his *Récréations mathématique et physiques* of 1696, a work subsequently edited and republished by Etienne Montucla nearly a century later. In Chapter 10, we find the following problem:

> To tell the number thought of by someone. Tell the person who has thought of a number to triple it and to take exactly half of the triple if it is even and the greater half if it is odd. Then tell him to triple that half and ask him how many times nine will go into the result. The number thought of will be twice that number of nines, plus one if it is odd.

For 11 and 12, the stages work out:

11	12
33	36
17	18
51	54
5	6
$11 = (2 \times 5) + 1$	$12 = 2 \times 6$

From all these examples we may perhaps appreciate the many varieties of such questions possible. Some are very simple, others can be quite subtle. Creating them is just a matter of a little simple algebra. We start with an unknown n or, if the individual digits are to be significant, with $10a + b$, $100a + 10b + c$, etc. We perform whatever operations we may choose and we arrange that the number reached at the conclusion of the calculations is either independent of the original number chosen or related to it in some specified way.

Magic Squares

A magic square of order n is a square array of n^2 whole numbers so arranged that the sums of every row, column and diagonal are equal. A magic square is said to be 'normal' when only the first n^2 numbers are used. If n is odd, the common sum is given by $n(n^2 + 1)/2$. Two examples of magic squares are

8	**1**	**6**
3	**5**	**7**
4	**9**	**2**

for which $n(n^2 + 1)/2 = 15$, and

23	**6**	**19**	**2**	**15**
10	**18**	**1**	**14**	**22**
17	**5**	**13**	**21**	**9**
4	**12**	**25**	**8**	**16**
11	**24**	**7**	**20**	**3**

for which $n(n^2 + 1)/2 = 65$.

We can construct a normal magic square of odd order very simply. The rule is:

Put **1** in the centre cell of the top row.
Put **2** and successive numbers up to n^2 diagonally North–East, allowing for edge-over adjacencies.
If the next cell according to this rule is occupied, use the cell immediately below the last-entered number.

If we follow this with the square of order 3 above, we begin with the **1** in the middle of the top row. Moving diagonally North–East takes us out of the square so we bring in the edge-over adjacency rule and put the next number, **2**, in the bottom right-hand cell. In other words, we imagine that one square stands on top of another and locate the cell for **2** accordingly. Moving diagonally North–East again takes us out of the square. This time the rule requires us to put **3** in the middle cell of the left-hand column – just as if another square were placed on the right. However, we now find the next cell that we want occupied by **1**, so **4** has to go immediately below **3**. When we have entered **6** in the top-right cell, the rule takes us to the cell occupied by **4**, so **7** must go below **6**. Further application of this rule take us to the cells for **8** and **9**.

In general, the method just described leads us to just one possibility out of several for any odd-order square. However, in the case of order 3 the alternative squares can all be obtained by rotation or reflection. The centre cell of a normal order-3 square is thus always occupied by **5**, and **1** can never occupy a corner cell.

The magic square of order 4 which appears in Albrecht Dürer's engraving *Melancholia*, is a particularly famous one and has a number of special properties. The rows, columns and diagonals all add up to 34. In addition, the four middle cells add up to 34, the four corner cells add up to 34, and so do the four top and bottom row interior cells and the four right-hand and left-hand column interior cells. This is by no means all, however. The sum of the eight numbers in the diagonals is equal to the sum of the remaining eight not in the diagonals. The sum of the squares of the numbers in the top two rows is equal

Magic square in
Dürer's *Melancholia*

to the sum of the squares of those in the bottom two rows, and also to the sum of the squares of those in the left two columns, and to the sum of the squares of these in the right two columns. The sum of the squares of the numbers in the diagonals is equal to the sum of the squares of the remaining numbers, and the same is true of their cubes. To crown all, Dürer arranged that the date of the engraving, 1514, should appear in the bottom row interior cells.

There is a rule for constructing normal magic squares of order $4n$. First, we consider the case where $n = 1$. We start by counting cells from left to right, row by row, and enter the numbers that coincide with non-diagonal cells only. We thus obtain

	2	**3**	
5			**8**
9			**12**
	14	**15**	

We now start from **1** again beginning at the bottom-right cell and working from right to left along each row until we finish at the top-left cell. This time we enter numbers coinciding with diagonal cells only. The completed square is then

16	**2**	**3**	**13**
5	**11**	**10**	**8**
9	**7**	**6**	**12**
4	**14**	**15**	**1**

This is almost the same as Dürer's square. The only difference is that the two interior columns are changed over.

We can construct a normal magic square of any order $4n$ on this principle by considering it to consist of n^2 sub-squares each of order 4. As we count from **1** to $\mathbf{16}n^2$ starting at the top-left cell, we fill in numbers in those cells which do not lie on the diagonal of any sub-square. Similarly, in recounting from the bottom-right cell, we enter all the numbers corresponding to cells on a diagonal of any sub-square. So the two stages of constructing a magic square of order 8 are

	2	**3**			**6**	**7**	
9			**12**	**13**			**16**
17			**20**	**21**			**24**
	26	**27**			**30**	**31**	
	34	**35**			**38**	**39**	
41			**44**	**45**			**48**
49			**52**	**53**			**56**
	58	**59**			**62**	**63**	

and

64	**2**	**3**	**61**	**60**	**6**	**7**	**57**
9	**55**	**54**	**12**	**13**	**51**	**50**	**16**
17	**47**	**46**	**20**	**21**	**43**	**42**	**24**
40	**26**	**27**	**37**	**36**	**30**	**31**	**33**
32	**34**	**35**	**29**	**28**	**38**	**39**	**25**
41	**23**	**22**	**44**	**45**	**19**	**18**	**48**
49	**15**	**14**	**52**	**53**	**11**	**10**	**56**
8	**58**	**59**	**5**	**4**	**62**	**63**	**1**

(The thick lines at the first stage identify the four sub-squares.) The common sum here is 260.

The oldest known magic square appears in the *I-king*, a Chinese work on permutations of uncertain date, possibly written by Won-Wang in the twelfth century B.C. Tradition has it that the *lo-shu*, as it is called, dates from the time of the Emperor Yu (*c.* 2200 B.C.). It is said that it decorated the back of a divine tortoise discovered on the banks of the Yellow River. The modern form of this square is

4	**9**	**2**
3	**5**	**7**
8	**1**	**6**

This magic square has had a long history in China, and even in this century it has been used all over the Far East by professional fortune tellers. It is still worn today, engraved on metal charms, as a protection against disease and evil spirits. It seems virtually certain that magic squares were a Chinese invention.

From China, the squares passed to Japan and to South–East Asia and also via India to the Arabs. They continued to be developed in China, however, where magic circles appeared also. We find discussions of these in the *Suan-fa Tong-tsung* (sixteenth century) and in the Japanese *Mantoku Jinkō-ki* (seventeenth century). The oldest magic squares from India date from the eleventh century or a little earlier, but it seems likely that they were known in India centuries before. An Arab work by Tâbit ibn Qorra, dating from the ninth century, deals with magic square properties. If, therefore, the theory of Chinese origin and dissemination via India is correct, they must have been known in India before the ninth century and probably much earlier.

As knowledge of magic squares travelled westwards, they became closely associated with both astrology and alchemy. They also occur in Hebrew cabalistic writings, where the order-3 square represents the forbidden name of God because the Hebrew alphabetical numerals representing fifteen are *yôdh* and *hē*, the first two letters of *Yahweh*. Normal magic squares of orders 3 to 9 were associated with the seven astrological heavenly bodies. The cells of the square of order 3 were also associated with metals and the square itself was thought to hold the secret which would turn base metals into gold – the dream of all alchemists. Although magic squares were never popular as Christian symbols, there are occasional references to them in mediaeval Latin writings. The square consisting of just one cell (containing the numeral **1**) was at one time thought to be a symbol of God, and, when it was discovered that a normal square of order 2 cannot be constructed, this was taken as symbolic of the devil or original sin. Magic squares continued, however, to be regarded as charms against disease and especially against the plague. They were engraved on silver talismans and sold widely throughout Europe during the Middle Ages.

Eventually, the squares began to be taken seriously by mathematicians and a number of works dealing with their mathematical properties began to appear. Emmanuel Moschopulus (fifteenth century) wrote a treatise on the theory of normal squares of odd order, much of which was included in a collection of theorems published in 1705 by Philippe de la Hire. De la Hire also edited the collected works of Bernard Frénicle, which included a mathematical study of magic squares. These works all discussed methods of construction. Sometimes, information was obtained directly from the East by western travellers. One of the envoys sent by King Louis XIV of France to Siam returned with details of methods of construction for normal squares acquired from Chinese living in the capital city. Cardano wrote on both the astrological significance and the mathematical properties of magic squares in his *Practica Arithmetica* of 1539. Throughout the next century and a half they were in considerable vogue throughout Europe and many works appeared in which they figured prominently.

One interesting development was the association of magic squares with the game of chess which had come to Europe from the Far East during the Arab incursions into Spain in the eighth century. The early methods of play differed significantly from those of today, and it was not until the sixteenth century that the modern game emerged, the last significant change being the introduction of castling (a technical term describing a particular move involving both the king and a castle of the same colour – the only occasion when moving two pieces at one turn is permitted). It is not certain who first investigated the relationship between the normal magic square of order 8 and the moves of a knight on a chessboard. Euler published a magic square in which each horizontal and vertical row adds up to 260, each horizontal and vertical row of the four sub-squares adds up to 130, and where a knight can make a complete tour of the board, occupying each cell once only, by following through the order of the numerals. This square is

1	48	31	50	33	16	63	18
30	51	46	3	62	19	14	35
47	2	49	32	15	34	17	64
52	29	4	45	20	61	36	13
5	44	25	56	9	40	21	60
28	53	8	41	24	57	12	37
43	6	55	26	39	10	59	22
54	27	42	7	58	23	38	11

In a sense it is not strictly a magic square as the diagonal sums do not equal the row and columns sums. This, however, seems only a small price to pay for the inclusion of the complete knight's tour.

The knight's move can be used for constructing magic squares of various order as has been known for several centuries. A very interesting eighteenth-century work by the Arab scholar, Muhammad ibn Muhammad, gives precise instructions for constructing order-5 squares by this method. We start with **1** in the top-right cell and then move two cells down and one to the left. This locates the cell in which we put **2**, and we then move a further two cells down and one to the left to locate the cell for **3**. This process is continued, making use of end-over adjacencies, to locate the cells for **4** and **5**, after which we find that the cell for **6** is already occupied by **1**. We then have to put **6** in the square two cells to the left of the **5** cell, and to continue knight's moves as before. At the end of each group of five numbers, we encounter an occupied cell and have to move two cells to the left instead of making a knight's move. The completed square is

13	**25**	**7**	**19**	**1**
17	**4**	**11**	**23**	**10**
21	**8**	**20**	**2**	**14**
5	**12**	**24**	**6**	**18**
9	**16**	**3**	**15**	**22**

The sum of each row, column and diagonal is 65, and so is the sum of every diagonal of five numbers taking account of end-over adjacencies. So, we have, for example

$$25 + 11 + 2 + 18 + 9 = 65$$

$$21 + 4 + 7 + 15 + 18 = 65$$

and so on.

Some of the modern elaborations of magic squares are quite remarkable and no doubt owe their existence to the computer. Squares of quite high order have been constructed, having all kinds of additional properties of which the knight's tour is only an elementary example. Additions can be made, not merely along rows, columns and diagonals, but also in blocks of cells of various shapes, within sub-squares, round the outsides of sub-squares, and so on. The possibilities seem endless, particularly if we do not restrict ourselves to the first n^2 whole numbers. Allowing negative numbers, and even fractions, provides opportunities for all sorts of ingenuity. The magic square principle is a gold-mine of recreational mathematics.

Number Shapes

The idea of representing positive whole numbers by dots arranged geometrically is a very old one, dating from the time of the early Pythagoreans (sixth century B.C.). Any positive integer can be regarded as a 'linear' number – we just represent it by a line of dots,

1 2 3 4 5

and so on. The squares of numbers can, of course, be represented by square arrays of dots:

1 4 9 16

and so on. Thus 1, 4, 9, 16, . . . , n^2 are 'square' numbers. We can form 'triangular' numbers from successive linear numbers:

1 3 6 10

and 'oblong' or 'rectangular' numbers from pairs of equal triangular numbers:

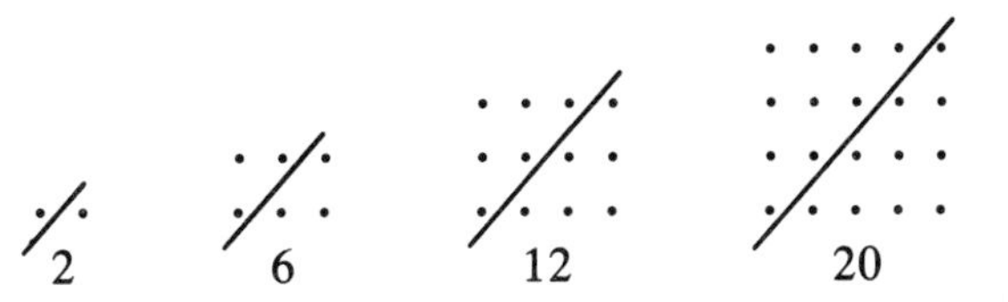

and so on.

Figurate numbers, as these patterns of dots are called, establish an important link between arithmetic and geometry. They are also extremely useful in establishing theorems about series of numbers.

The linear number L_n is just n itself, thus $L_1=1$, $L_2=2$, $L_3=3$, and so on. The triangular number T_n is the sum of the first n linear numbers, that is,

$$\begin{aligned} T_n &= L_1 + L_2 + \ldots + L_n \\ &= 1 + 2 + \ldots + n \end{aligned}$$

We can obtain a formula for this sum by reference to rectangular numbers. Each triangular number is half the corresponding rectangular number, and each rectangular number R_n is just $n(n + 1)$. So

$$T_n = \tfrac{1}{2}n(n + 1)$$

We have thus obtained an expression for the sum of the first n numbers.

If we look at any square number S_n, say $S_4 = 16$, we can see that there are two ways in which this can be built up apart from just drawing a square array of dots. We can form it by the addition of the two triangular numbers T_3 and T_4 as follows:

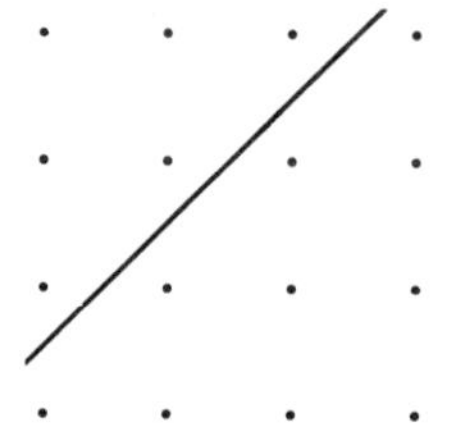

so that we have

$$S_4 = 16 = 6 + 10 = T_3 + T_4$$

This is true generally, so that

$$S_n = T_{n-1} + T_n$$

which just expresses the identity

$$n^2 = \tfrac{1}{2}n(n-1) + \tfrac{1}{2}n(n+1)$$

Alternatively, we can picture S_4 built up successively as

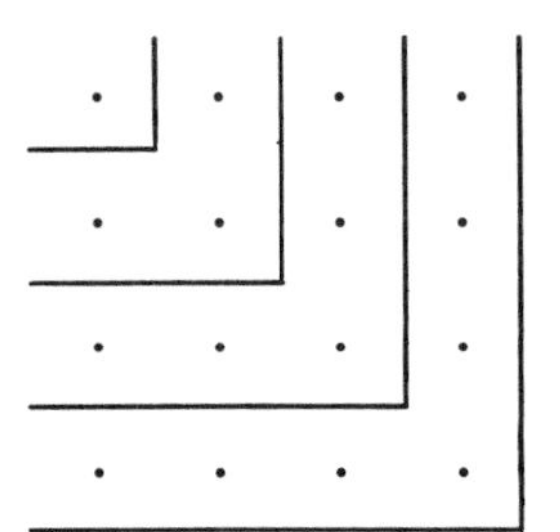

These reversed–L shapes are known as 'gnomons', and they contain just the successive odd numbers, 1, 3, 5, . . . , $(2n - 1)$. This shows us that n^2 is the sum of the first n odd numbers. We can also build up rectangular numbers in this way, so that, for example, $R_5 = 30$ is seen as the sum of the first five even numbers:

In general,

$$R_n = 2 + 4 + 6 + \ldots + 2n$$
$$= n(n+1)$$

We can also show quite easily that the square of any odd number is one more than the sum of eight identical triangular numbers. Algebraically, we have

$$\begin{aligned}(2n + 1)^2 &= 4n^2 + 4n + 1 \\ &= 4n(n + 1) + 1 \\ &= 8[\tfrac{1}{2}n(n + 1)] + 1 \\ &= 8T_n + 1\end{aligned}$$

The corresponding picture given by figurate numbers in the case $n = 2$, is

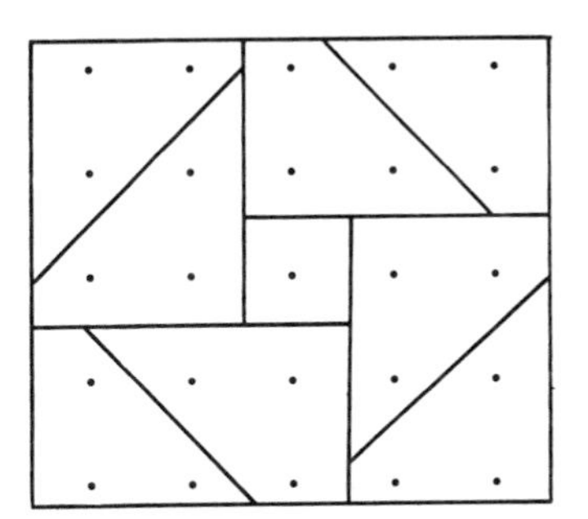

The odd unit always appears in the centre of the figure.

Plane figurate numbers can be extended to include 'pentagonal' numbers, 'hexagonal numbers', and so on, and many more numerical relationships can easily be demonstrated. The *n*th pentagonal number, given by

$$1 + 4 + 7 + \ldots + (3n - 2)$$

is the sum of L_n and three T_{n-1}s. We therefore have

$$P_n = L_n + 3T_{n-1}$$

that is,

$$1 + 4 + 7 + \ldots + (3n - 2) = n + \tfrac{3}{2}n(n - 1) = \frac{n(3n - 1)}{2}$$

Thus, for P_4, we have

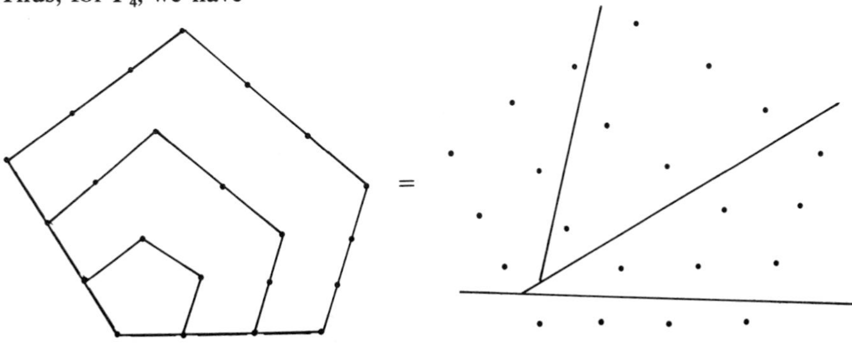

In three dimensions, we have many possibilities open to us. The most obvious ones are cubes, and triangular and square pyramids. A cube is just represented by a three dimensional array of n^3 dots. Thus C_3 is

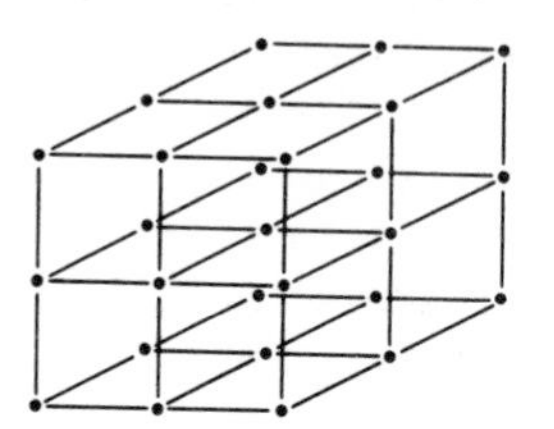

It is built up out of n layers of S_n. A triangular pyramidical number is formed by successive layers of $T_1, T_2, T_3, \ldots, T_n$. Thus

$$TP_n = T_1 + T_2 + \ldots + T_n$$

and so

$$TP_3 = T_1 + T_2 + T_3 = 1 + 3 + 6 = 10$$

Its figure is

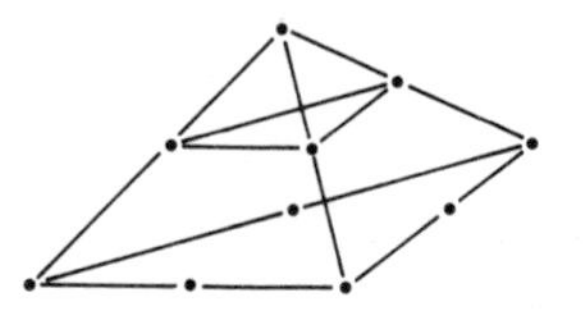

Similarly, a square pyramidical number is formed by successive layers of S_1, $S_2, S_3, \ldots, S_n$. Thus,

$$SP_n = S_1 + S_2 + \ldots + S_n$$

and so

$$SP_3 = S_1 + S_2 + S_3 = 1 + 4 + 9 = 14$$

Its figure is

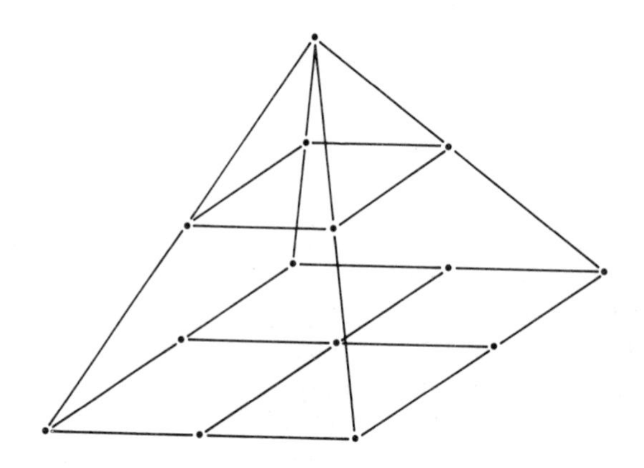

It is useful to make a table of the values of figurate numbers, as the table then suggest further relationships between them. The complete table up to $n = 5$ is

n	L_n	T_n	S_n	R_n	TP_n	SP_n	C_n
1	1	1	1	2	1	1	1
2	2	3	4	6	4	5	8
3	3	6	9	12	10	14	27
4	4	10	16	20	20	30	64
5	5	15	25	30	35	55	125

From the table, we can see the relations:

$$S_n + C_n = nR_n$$

$$T_n + S_n + C_n = 3SP_n$$

and from these derive the usual formula for the sum of the first n square numbers. Thus we have

$$\begin{aligned} T_n + S_n + C_n &= \tfrac{1}{2}n(n + 1) + n^2(n + 1) \\ &= \tfrac{1}{2}n(n + 1)(2n + 1) \end{aligned}$$

so that

$$SP_n = \tfrac{1}{6}n(n + 1)(2n + 1)$$

We can also see from the table that the sum of the first two cubes, $1 + 8 = 9$, is the square of the second triangular number, 3, and that generally

$$C_1 + C_2 + \ldots + C_n = T_n^3$$

This gives us the usual formula for the sum of the first n cubes, $[\tfrac{1}{2}n(n + 1)]^2$.

Much of this was known to the Pythagoreans, though, curiously enough, figurate numbers do not appear in Euclid's works. The surviving Greek works which treat figurate numbers in detail belong to the neo-Pythagorean period of Theon of Smyrna and Nicomachus of Gerasa (second century A.D.). In his *Introduction to Arithmetic*, Nicomachuś developed many of the number relations of the Pythagoreans, including the relation between the sums of succeeding odd numbers and the cubes expressed as

$$\underbrace{1}_{1^3},\ \underbrace{3 + 5}_{2^3},\ \underbrace{7 + 9 + 11}_{3^3},\ \underbrace{13 + 15 + 17 + 19}_{4^3},\ \ldots$$

All this reappeared in the *De institutione arithmetica* of Boethius (sixth century), through which figurate numbers eventually became known in western Europe.

This kind of geometrical representation of whole numbers had a number of practical uses. It was used by Roman surveyors for the approxiamte calculation of areas. As early as the third century B.C., it was used by Archimedes in the development of his methods for calculating areas bounded by curves, and it continued to be associated with this topic until the seventeenth century. It influenced the way in which Greek mathematicians thought about numbers. However, it was sometimes taken to ridiculous lengths. Attempts were made to calculate numerical values to be associated with animals whose shapes were drawn using vast numbers of dots. Thus, a horse might be given the value 376, simply because of a picture of it drawn entirely in dots. This kind of application of figurate numbers lasted only for a few centuries, and by late Renaissance times interest in them, both mathematical and speculative, decreased rapidly. They have reappeared in modern times only in connection with the teaching of rudimentary mathematical ideas to the very young, where they are of considerable educational value.

As we have seen, figurate numbers normally involve patterns of dots, though in some texts we find patterns of small squares or diamonds, or even of the Greek letter α. It has been suggested that this was just the first letter of *arithmos*, but it seems more likely that it was the alphabetical numeral for one, since the dot was a symbol of the basic indivisible unit of arithmetic and each dot in a figurate number had the value unity. A dot and the letter α were thus equivalent.

There is one particularly famous array of numbers in a geometrical shape; this is the triangular array

$$
\begin{array}{ccccccccccc}
 & & & & & 1 & & & & & \\
 & & & & 1 & & 1 & & & & \\
 & & & 1 & & 2 & & 1 & & & \\
 & & 1 & & 3 & & 3 & & 1 & & \\
 & 1 & & 4 & & 6 & & 4 & & 1 & \\
1 & & 5 & & 10 & & 10 & & 5 & & 1 \\
\cdot & \cdot & \cdot & \cdot & \cdot & \cdot & \cdot & \cdot & \cdot & \cdot & \cdot
\end{array}
$$

usually called 'Pascal's triangle'. In fact it is much older than Pascal's time (seventeenth century), but it became associated with his name on account of his extensive discussion of its properties in the *Traité de Triangle Arithmétique*, published posthumously in 1665.

If we look at the entries in the triangle, it is not difficult to see that each is the sum of the two immediately above it. Thus $2 = 1 + 1$, $3 = 1 + 2$, $4 = 1 + 3$, $6 = 3 + 3$, and so on. Also, the sum of all the numbers in any row is twice the sum of those in the row immediately above.

Pascal's triangle is associated particularly with coefficients of the expansion of $(a + b)^n$ and with elementary probability. If we write out successive binomial (two-number) expansions, we have

$$
\begin{aligned}
(a+b)^0 &= 1\\
(a+b)^1 &= 1a + 1b\\
(a+b)^2 &= 1a^2 + 2ab + 1b^2\\
(a+b)^3 &= 1a^3 + 3a^2b + 3ab^2 + 1b^3\\
(a+b)^4 &= 1a^4 + 4a^3b + 6a^2b^2 + 4ab^3 + 1b^4\\
(a+b)^5 &= 1a^5 + 5a^4b + 10a^3b^2 + 10a^2b^3 + 5ab^4 + 1b^5\\
\ldots & \quad \ldots \quad \ldots \quad \ldots \quad \ldots
\end{aligned}
$$

The coefficients of the various terms are exactly those which appear in the triangle. Thus, we can immediately write out the expansion of $a + b$ to any power simply by reference to the appropriate row of the triangle. Since $a + b$ is a symmetric expression, the triangle is symmetrical about a vertical line through its apex – the entries on the right are a mirror-image of those on the left. We can, however, use the triangle to expand the more general binomial $(ma + nb)^n$ simply by treating ma as a and nb as b. Thus:

$$(2a + 3b)^3 = 1(2a)^3 + 3(2a)^2(3b) + 3(2a)(3b)^2 + 1(3b)^3$$
$$= 8a^3 + 36a^2b + 54ab^2 + 27b^3$$

It was the expansion of binomial expressions which first led to Pascal's triangle.

To understand the association of the triangle with probability theory, we should consider what happens when we toss a number of coins. If we toss one coin, then we have a fifty-fifty chance of getting one of the two possible outcomes, say 'heads'. If we toss two coins, then there are four equally likely outcomes:

HH HT TH TT

If we toss three coins, there are eight equally-likely outcomes:

HHH HHT HTH THH
HTT THT TTH TTT

Now let us look at the number of times which 'tails', say, appears in these various outcomes. We can tabulate this as follows:

No. of coins	0T	1T	2T	3T
1	1	1		
2	1	2	1	
3	1	3	3	1

And we can continue this process as far as we like – we just obtain the numbers in succeeding rows of Pascal's triangle. This means that we can read off probabilities directly from the triangle. The total number of possible outcomes is the sum of the appropriate row; the number of times we would expect a specified outcome is given by the individual entry in that row. The probability of getting two 'tails' if we throw three coins is 'three times in eight goes', that is $\frac{3}{8}$. Similarly, by referring to the sixth row, whose sum is 32, the probability of getting two 'tails' on throwing five coins is $\frac{5}{32}$, and so on. The 1 at the apex can be interpreted as the probability of getting no 'tails' if we do not throw any coins.

All that we have written about tossing coins would apply equally well to any similar situation – the number of boys and girls in a family, for example, assuming that a baby has a fifty-fifty chance of being a boy (or a girl).

An alternative way of writing down the Pascal triangle is:

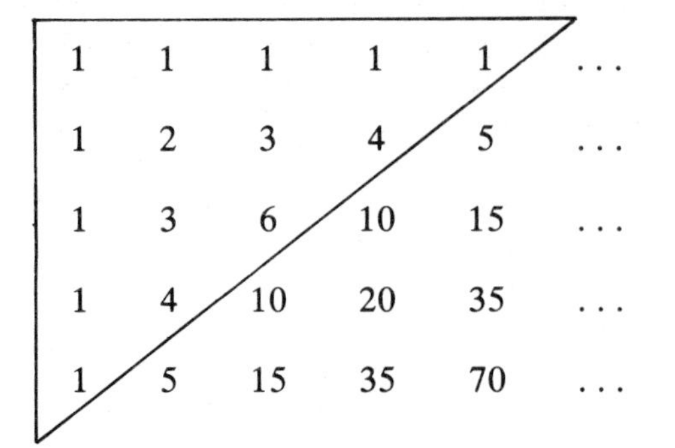

This is, in fact, the way in which Pascal himself presented it. In this form, it is easier to see further relations between the numbers. For example, each number in the triangle is the sum of the numbers which lie above and to the left in the row immediately above. Thus

$$10 = 1 + 3 + 6$$

and

$$35 = 1 + 4 + 10 + 20$$

Each number is also one more than the sum of the numbers in the rectangle (or square) above and to the left of it. Thus

$$6 = (1 + 1 + 1 + 2) + 1$$
$$10 = (1 + 1 + 1 + 2 + 1 + 3) + 1$$

and

$$20 = (1 + 1 + 1 + 1 + 2 + 3 + 1 + 3 + 6) + 1$$

There are many other relations to be found.

The earliest known occurrences of Pascal's triangle come from China. It appears in the writings of Yang Hui (thirteenth century) with entries up to the seventh row, that is, giving the binomial coefficients of the expansion of $a + b$ up to its sixth power. It also appears in the better-known *Ssu Yuan Yü Chien* of Chu Shîh-Chieh (1303), where it goes as far as the ninth row. There are references in these and other Chinese works which suggest that the triangle was known in the eleventh century. It seems certain, therefore, that it is of Chinese origin.

The triangle appears in a number of Arab works from the twelfth century onwards. Its first appearance in print in Europe is to be found on the title page of Peter Apian's *Rechnung* of 1527. After this date, it is to be found in works by many different authors. It appears in Tartaglia's *General trattato di numeri e misure* of 1556, where the author claims it to be his own invention. It also appeared in the sixteenth century in works by Stifel, Scheubel and Bombelli, and in a number of seventeenth-century works before 1665.

All the early occurrences of the triangle are associated with binomial expansions; it was Pascal who associated it specifically with probabilities. He is therefore often given the credit for founding probability theory. This also is hardly appropriate since games of chance had been discussed at some length a century earlier in Cardano's *Liber de Ludo Aleae*. However, although we must challenge the attribution to Pascal both of the invention of the triangle and of the founding of probability theory, we must not underestimate the value of the study of all aspects of the triangle which appears in his *Traité* of 1665. As with the case of magic squares, what may have first started as a kind of mathematical recreation eventually came to be the subject of serious mathematical investigation, and Pascal's work stands out from amongst all the others for its elegant and systematic thoroughness.

Numbers with Special Properties

The Pythagoreans were particularly interested in properties of whole numbers. Their interest was a mixture of mathematics and mysticism. Certain numbers, for example, were regarded as 'perfect', others as 'friendly'.

It would seem that in the first instance, 10 was regarded as perfect because it was the sum of the numbers 1 to 4, each of which had a special mystical significance. However, the later Pythagoreans defined a perfect number to be one which was the sum of its proper divisors, that is, the sum of all numbers (including unity but excluding the number itself) which divide exactly into it. If we neglect 1, the first perfect number is 6. Its proper divisors are 1, 2 and 3, and $1 + 2 + 3 = 6$. The next perfect number is 28. Its proper divisors are 1, 2, 4, 7 and 14, and $1 + 2 + 4 + 7 + 14 = 28$. A chapter of Euclid's *Elements* includes a discussion of these numbers, and there is a proof of the theorem 'if 2^{n-1} is a prime number then $2^{n-1}(2^n - 1)$ is perfect'. We find further discussions of perfect numbers in the writings of Nicomachus, who quotes 496 and 8128 as being perfect in addition to 6 and 28. All the numbers given by this formula must be even. After 8128, the next is 33 550 336. We do not know if any odd perfect numbers exist. As recently as 1952, there were only twelve perfect numbers known. The development of computers has now considerably extended the list. Just how many there are is not known, nor do we have any way of representing their distribution amongst the whole numbers.

It is said that when Pythagoras was asked to describe the characteristics of a friend, he replied 'he is the other I, like 220 and 284'. From this tradition, such pairs of number came to be called 'friendly' or 'amicable'. The property which they have is that each is the sum of the proper divisors of the other. The proper divisors of 220 are 1, 2, 4, 5, 10, 11, 20, 22, 44, 55 and 110. These add up to 284. Similarly, the proper divisors of 284, which are 1, 2, 4, 71, 142, add up to 220. The special relationship of these two numbers was thought to have all kinds of magical properties. Friends exchanged talismans engraved with these numbers; men, whose names could be made equal to one of them by gematria, sought brides with names equivalent to the other. They are mentioned in many Arab works, and discussed by Renaissance writers such as Chuquet, Pacioli and the friar Mersenne.

The Greeks were apparently unable to discover other pairs of friendly numbers, and it is for this reason that 220 and 284 came to have such special significance. However, in 1636 Fermat calculated a second pair, 17 296 and 18 416, probably after studying the works of Tâbit ibn-Qorra (ninth century). Two years later, Descartes found the pair, 9 363 584 and 9 437 056, and Euler extended the number of known pairs to sixty. Towards the end of the nineteenth century, Nicolo Paganini discovered that 1184 and 1210 are perfect. Whether this was the first time that this particular pair was known is open to question. It seems hardly likely that a pair of such comparatively small numbers could have been overlooked over the centuries, particularly as we know that there had been extensive searches for such pairs from Greek times onwards. However, Paganini was only sixteen and it is highly unlikely that he would have access to historical documents totally unknown to other mathematicians. Whether or not his particular pair had been discovered earlier, his achievement was remarkable and must have involved a considerable amount of hard work, since he worked entirely on a trial-and-error basis. The use of computers is steadily extending the list of known friendly numbers, but we still have no general theory which will account for their distribution.

The numbers 9 and 11 have interesting properties because they are respectively one less and one more than ten. Some of these properties were used for checking arithmetic calculations. If we examine the squares of those numbers which consist just of nines, we find:

$$9^2 = 81$$
$$99^2 = 9801$$
$$999^2 = 998\ 001$$
$$9999^2 = 9\ 998\ 001$$

and from this we infer that

$$99\ 999^2 = 9\ 999\ 800\ 001$$

We just write a further 9 at the beginning of the number and insert another 0 before the final 1. We have a similar situation with cubes of numbers consisting just of nines:

$$9^3 = 729$$
$$99^3 = 970\ 299$$
$$999^3 = 997\ 002\ 999$$
$$9999^3 = 999\ 700\ 029\ 999$$

We infer that

$$99\ 999^3 = 999\ 970\ 000\ 299\ 999$$

Here we add a 9 at each end and insert an additional 0 in the middle. For the fourth powers, we do not encounter a similar situation.

Another property of 9 arises from division. If we divide the numbers 1 to 8 successively by 9, we obtain:

$$1 \div 9 = 0{\cdot}\dot{1} \qquad \text{(that is, } {\cdot}111\ 111 \ldots)$$
$$2 \div 9 = 0{\cdot}\dot{2} \qquad \text{(that is, } {\cdot}222\ 222 \ldots)$$
$$3 \div 9 = 0{\cdot}\dot{3}$$
$$\ldots \qquad \ldots$$
$$8 \div 9 = 0{\cdot}\dot{8}$$

If we now divide 9 by 9, we just get 1, but in fact this is the same number as $0{\cdot}\dot{9}$.

If we take any three-digit number for which there are more tens than units and more hundreds than tens, reverse the number and subtract, we will always obtain another three-digit number with nine tens, and with the sum of the other two numbers equal to nine. Thus

$$852 - 258 = 594$$

and

$$631 - 136 = 495$$

The number 11 is associated with numerical palindromes – numbers which read the same from right to left as from left to right. We have

$$11$$
$$11^2 = 121$$
$$11^3 = 1331$$
$$11^4 = 14641$$

These numbers are formed from the digits of successive rows of Pascal's triangle and are accordingly all palindromes. We might be tempted to expect that 11^5 would also give us a palindrome, but this is not so. We can, however, write down the fifth power of 11 directly from the next row of the triangle simply by carrying the tens figure of 10 to the next place. This gives us

$$11^5 = 161\ 051$$

Similarly, by carrying the tens figures of 20 and the two fifteens, we can immediately write down

$$11^6 = 1\ 771\ 561$$

and so on. We are just carrying out successive binomial expansions of $(10 + 1)^n$.

Palindromes can be obtained by squaring the numbers 1, 11, 111, 1111, and so on. This gives us:

$$1^2 = 1$$
$$11^2 = 121$$
$$111^2 = 12321$$
$$1111^2 = 1234321$$
$$\ldots \qquad \ldots$$

Eventually we have:

$$111\,111\,111^2 = 12345678987654321$$

A palindrome can also be obtained by choosing any number, reversing it, adding, then reversing and adding again, and so on. For example, if we choose 3821 reverse it and add, we get

$$\begin{array}{r} 3821 \\ 1283 \\ \hline 5104 \end{array}$$

Reversing 5104 and adding gives us

$$\begin{array}{r} 5104 \\ 4015 \\ \hline 9119 \end{array}$$

which is a palindrome. We could now add 9119 to itself, giving 18 238. Reversing and adding gives

$$\begin{array}{r} 18238 \\ 83281 \\ \hline 101519 \end{array}$$

almost, but not quite another palindrome. This will work if all the digits of the chosen number are less than 5, or if the sums of the first and last, second and next-to-last, etc. digits are not greater than 9, for we simply have (for a four-figure number):

$$\begin{array}{ccccccc} 1000a & + & 100b & + & 10c & + & d \\ 1000d & + & 100c & + & 10b & + & a \\ \hline 1000(a+d) & + & 100(b+c) & + & 10(b+c) & + & (a+d) \end{array}$$

When a carry is involved, we just carry on until these favourable conditions apply.

Sometimes we find that carrying out a given succession of operations on arbitrarily chosen numbers always gives us the same result. Consider the following sequence of operations:

Take any three-digit number with a descending order of digits.
Reverse it.
Subtract.
Reverse the result.
Add.

Surprisingly, we invariably get 1089. Two examples are

$$\begin{array}{r} 862 \\ 268 \\ \hline 594 \\ 495 \\ \hline 1089 \end{array} \qquad \begin{array}{r} 941 \\ 149 \\ \hline 792 \\ 297 \\ \hline 1089 \end{array}$$

To see why this happens, suppose that our chosen three-digit number is $100a + 10b + c$. We can write this as $100(a - 1) + 10(10 + b - 1) + (10 + c)$. Now we subtract $100c + 10b + c$ and obtain $100(a - c - 1) + 10(9) + (10 + c - a)$. We can now see just why reversing and subtracting always gives us a 9 in the middle and the hundreds and units digits summing to 9. We have to reverse this and add, that is, we add $100(10 + c - a) + 10(9) + (a - c - 1)$. This gives us $100(9) + 10(18) + 9$, which is just 1089.

The number 1089 has other properties. If we multiply it by 9 we get 9801, that is, we reverse the digits. If we multiply it by 2, we get 2178, and 4 times 2178 is 8712; again we reverse the digits. If we multiply it by 4 we get 4356, and $1\frac{1}{2}$-times 4356 is 6534. It we multiply it by 5, we get the palindrome 5445. These properties can all be explained, there is nothing magical about them, though they can be used to mystify the uninitiated. Significantly, $1089 = 9 \times 11 \times 11$ and $9801 = 9 \times 9 \times 11 \times 11$, so both 9 and 11 are involved here.

The number 7 is associated with sets of cyclic numbers. The fractions $\frac{1}{7}, \frac{2}{7}, \ldots, \frac{6}{7}$, when expressed as recurring decimals are

$$\cdot 142\,857 \;\; 142\,857 \ldots$$
$$\cdot 285\,714 \;\; 285\,714 \ldots$$
$$\cdot 428\,571 \;\; 428\,571 \ldots$$
$$\cdot 571\,428 \;\; 571\,428 \ldots$$
$$\cdot 857\,142 \;\; 857\,142 \ldots$$

We see that in each case the cyclic order of the digits is retained. The digits 3, 6 and 9 are missing, but have we escaped from the all-pervading 9? If we multiply 142 857 by 7 we get 999 999, so we are not as far away from 9 as we might have supposed! Indeed, if we multiply the recurring fraction

$$\cdot 142\,857\,142\,857 \ldots\ldots$$

by 7 we get

$$\cdot 999\,999\,999\,999 \ldots\ldots$$

This is, however, only to be expected since we are multiplying $\frac{1}{7}$ by 7 and, as we have already remarked, $\cdot\dot{9}$ is the same number as 1.

In Chapter 5, we briefly discussed binary numbers, that is, numbers written in powers of 2 and hence using only the numerals 0 and 1. If we take four plain cards and write on the first (in decimal notation) all the numbers up to 15 whose binary representation ends in 1, on the second all those whose binary representation has a 1 in the two's place, on the third all those with a 1 in the two-squared place, and finally on the fourth all those with a 1 in the two-cubed place, we shall have:

Card 4		*Card 3*		*Card 2*		*Card 1*	
8	12	4	12	2	10	1	9
9	13	5	13	3	11	3	11
10	14	6	14	6	14	5	13
11	15	7	15	7	15	7	15

We can now ask a friend to think of any number from 1 to 15, remember the cards on which it occurs, and then turn the cards over. On being told which cards contain the chosen number, we can name this number immediately. If, for example, 14 is chosen, we shall be told that it is in the second, third and fourth cards. We immediately identify the chosen number in binary form as 1110, which is 14 in decimal notation. Similarly, if 9 is chosen, the first and fourth cards are indicated, from which we deduce $1001 = 9$. We can do the same thing with 5 cards for numbers up to $2^5 - 1 = 31$, with 6 cards for numbers up to $2^6 - 1 = 63$, and so on. The more cards we use, the more easily will it seem that we have not memorized the numbers on them.

The possibilities for recreation inherent in the whole numbers are limitless. It is not at all surprising that the layman, just as much as the professional mathematician, has devoted much time and patience to inventing number problems and puzzles of all kinds. Number theory has been a part of formal mathematical activity since the beginning of the Pythagorean period at least, and probably much earlier. It is also, however, a part of informal mathematical activity and has given pleasure to a great many people and will continue to do so. It is an area in which the non-specialist can make noteworthy contributions – witness the two friendly numbers discovered by Paganini. Many of the interesting problems of number theory have appeared, not in professional mathematical journals but in journals of mathematical recreation, in teachers' magazines, and in publications of a completely general character. Number problems have been especially popular with the armed forces in wartime, and they can be a valuable interest for those who find themselves confined to bed or even in prison. It seems that they can have a therapeutic value in all sorts of unexpected environments.

Chapter Eight

Thinking about Numbers

It is not possible that without numbers
anything can be either conceived
or known.

(Philolaus)

God created the integers, the rest
is the work of man.

(Kronecker)

We have seen that man's first appreciation of numbers was related to their application in everyday life. Any particular number was a property possessed by specific collections of objects. Thus 'being two' was common to the man–woman pair, to a person's arms or legs or eyes, and so on. Similarly, the property of 'being three' was common to, perhaps, the children in a family, the sacks of corn in a barn, a particular group of trees, and so on. Support for the theory that numbers were first thought of in this way comes from surviving adjectival forms of the first few number words.

We have also seen that, before counting can take place, there must be both a cardinal and an ordinal understanding of numbers. Man had to appreciate that numbers answer not only 'how many?' but also 'where in a prescribed order?' There are good reasons for believing in the priority of the latter. The head man of a village community was first, just as the husband was first in the family. But being first, or second, or third was an attribute of particular people or objects in particular contexts. At what stage these attributes came to have an existence of their own, we cannot say. It seems more likely that it was via the cardinal numbers that numbers were first thought of as entities in their own right. This may have come about because it is these that are manipulated in elementary calculation. That two sacks and three sacks make five sacks and that two sheep and three sheep make five sheep are facts more likely to give birth to the abstract idea that two and three make five than that being third after the second makes something or someone the fifth. This is, however, a speculative approach without serious historical basis.

When we come to the ancient civilizations about which we do have concrete evidence, it is clear that there was already considerable skill in the manipulation of numbers. A few of the problems described in the earliest texts do suggest that not all the calculations were strictly practical; some indeed seem to be carried out as abstract operations for educational or recreational purposes. We also have widespread evidence of the association of numbers with traditional beliefs. This does seem to suggest that numbers had come in some way to have an existence of their own, though certainly not in as abstract a way as they came to have for the Pythagoreans and later Greek mathematicians.

We can be sure, however, that for all ancient peoples numbers essentially meant positive whole numbers, or, as we call them, 'natural' numbers. These were sufficient for their counting and measuring. Numbers could be chosen as large as was needed to meet any practical eventuality; they did not run out when a large army was being counted, though it was possible to run out of symbols. Fractions were used, but the evidence suggests that they were often avoided – usually by making use of smaller units of measurement. Two-thirds, for example, was the most common of the fractions not having unity as the numerator and was not thought of as a part of a whole, but as two third-parts, one third-part being then the basic unit.

Until Greek times there was no special distinction made between the discrete and the continuous. For all practical purposes the discrete was adequate for measuring the continuous; a length could be measured to any degree of accuracy required. The Babylonians could calculate the square root of two to considerable accuracy. There was no reason to suspect that numbers, as they understood them, were incapable of measuring the diagonal of a unit square. In the same way, the perimeter of a circular stone of known radius could be measured as accurately as desired. There was no reason to suspect that the relationship of the perimeter to the radius could not be expressed in terms of their numbers. It was left to the Greeks to stumble upon the problems inherent in their conceptual approach to numbers, which required abstract proofs rather than practical verifications.

Rational and Irrational Numbers

For the Pythagoreans, the natural numbers, 1, 2, 3, . . . , lay at the heart of the universe, and they found this belief supported by their experience of the world. Discovery of the way in which musical sounds corresponded to proportions between numbers and the relationship of numbers to geometry via figurate numbers strengthened their belief that numbers could satisfactorily explain the entire physical world. The geometric point was thought of as a basic unit corresponding to the unity of the number sequence. Lines then became strings of points, areas were built up from lines and solids from areas.

We know very little about the transition from the largely practical approach of Babylonian and Egyptian mathematics to the largely religio-philosophical approach of the early Greeks. We know that Babylonian and Egyptian mathematics was transmitted to the Greek culture, and we know what the Greeks eventually created by fusing it with their own ideas; but the detailed process of this fusion and the precise cultural circumstances in which it took place are unknown. What little we know is based on tradition and on remarks by later Greek commentators whose assessment of the period concerned is known not to be entirely reliable.

Pythagoras himself lived in the sixth century B.C. It is said that he was a pupil of the even more misty figure, Thales, and that he founded an elitist religious brotherhood which held property in common and followed strict rules of diet and physical exercise. Their studies covered the whole of human life, including politics, and it was their political activity which is said to have led eventually to the destruction of their community at Croton and the death of Pythagoras himself. It is impossible to say how much of Pythagorean mathematics was due specifically to the head of the community. The practice in those days was to ascribe all the mathematical, scientific and philosophical ideas of a school to its leader. We thus have no way of distinguishing between the discoveries of Pythagoras himself and those of his many pupils.

The natural numbers and their ratios provided the Pythagoreans with an adequate basis for all their mathematical activity until it was discovered that no ratio of numbers can define the length of the diagonal of a square in terms of the length of its sides. This difficulty was bound to arise out of the well-known geometrical theorem (attributed to Pythagoras) that the square on the hypotenuse of a right-angled triangle is equal to the sum of the squares on the other two sides. Pythagorean triples had been familiar to the Babylonians long before the time of Pythagoras as is witnessed by the tablet, Plimpton 322. There is no reason to suppose, however, that the Babylonians had any general proof of Pythagoras' theorem. It is therefore possible that the first proof was due to Pythagorus himself, though its actual form probably differed from the traditional classroom proof found in Euclid's *Elements* (Book I, Proposition 47).

We cannot be certain how the incommensurability of the diagonal and sides of a unit square was first proved. According to Aristotle, the proof depended upon the distinction between odd and even numbers. The traditional Greek proof is a model of proof by *reductio ad absurdum*.

According to Pythagoras' theorem, the square on the diagonal of the unit square must have area 2. The length of the diagonal is therefore that number whose square is 2. If it is commensurable with the length of the sides, we shall be able to represent it by a ratio, that is, it will be expressible as p/q where p and q are natural numbers. But ratios are not unique;

$$\frac{ap}{aq} = \frac{p}{q}$$

for any number a. We can always assume, however, without loss of generality that any factors common to both p and q have been cancelled out, that is, p/q is in its lowest terms. Now, since the length of the hypotenuse is p/q, the area of its square must be p^2/q^2, which in turn must be 2. If

$$\frac{p^2}{q^2} = 2$$

then

$$p^2 = 2q^2$$

so p^2 must be even, and p must in turn be even. If p is even, we can write it as $2r$ and p^2 as $4r^2$. We therefore have:

$$4r^2 = 2q^2$$

and (on dividing by 2)

$$2r^2 = q^2$$

so q^2 and hence q are even. But, if p and q are both even, p/q is not in its lowest terms, and this contradicts our earlier assumption that the length of the hypotenuse can be expressed in the form p/q, where this ratio is in its lowest terms. Since our assumption has led to a contradiction, it cannot be true – that is, the length of the hypotenuse must be incommensurable with the sides, or, in purely numerical terms, $\sqrt{2}$ is irrational.

We should note the form of this beautiful proof. The *reductio ad absurdum* method begins with the assumption that whatever we want to prove is false. Thus, we assumed that the diagonal is commensurable with the sides and that its length can therefore be expressed in the form p/q where this ratio is in its lowest terms – that is, we assumed its incommensurability to be false. The method then requires the deduction of a contradiction from the assumption, that is, the deduction of something which either contradicts the assumption itself or is clearly untrue for other reasons. In our example we deduced that p/q could not be in its lowest terms, a clear contradiction of our original assumption. The final stage argues that since the assumption has led to a contradiction, it must be false. Here, this means that the assumption of commensurability is false, and the proof is thus complete.

Proof by *reductio ad absurdum* is a Greek invention and plays a very important role in Greek mathematics. It has been challenged on the grounds that it is negative rather than positive, and because it relies on the assumption that two falsehoods can lead us to truth. Most mathematicians do, however, accept such proofs, largely because to reject them considerably limits the total amount of proved mathematics. Many theorems, usually proved by *reductio ad absurdum* can be proved by other methods, though often only at considerable length. Some generally accepted theorems cannot be proved otherwise, however, and to the majority of mathematicians there seems little point in rejecting these – and hence the mathematics which depends on them – for the sake of what they would regard as a logical nicety.

We do not know who was the first to discover irrational numbers. Tradition suggests that it may have been a certain Hippasus. There are, however, various conflicting accounts both about what Hippasus did to anger the rest of the Pythagorean brotherhood and about what eventually happened to him. It is alleged that for revealing the discovery that there were lengths which could be constructed but which could not be measured by ratios of numbers, he was drowned at sea. Other accounts simply say that he was expelled for leading an internal revolt against the strict rules of the brotherhood. Who actually made the discovery of irrationality does not matter all that much – it is the effects of it which are important in the history of numbers, especially as these effects were reinforced by the activities of a rival philosophical school, followers of Parmenides of Elea.

The Eleatic school attacked the numerical atomism of the Pythagoreans. They used what was to become the Socratic method, that is, by assuming their opponents tenets they followed through arguments based on these which led to absurd conclusions. The best-known of these arguments are those of Zeno (about 450 B.C.). He argued that if space and time are made up of strings of ultimate indivisible elements, points and instants, then motion is impossible. For, if the tip of an arrow occupies a given point in space at a given instant of time, it cannot both be occupying that point and moving from that point at the given instant. There is thus no way in which the tip of the arrow can occupy the next point in space at the next instant of time. This holds for all points and all instants; hence the arrow cannot be moving. A similar argument was put forward in another of Zeno's paradoxes on motion known as the *Stade*. The conclusion was again that if space and time are made up of ultimate indivisible elements motion is impossible.

The effect of such arguments was to reinforce the incommensurable difficulties arising from the assumption that all lengths could be expressed in terms of ratios of natural numbers. This has often been described as a great crisis in the evolution of Greek mathematics. It is, however, probably something of an exaggeration to suggest that the mathematics of the Pythagoreans was thrown entirely into disarray. It is much more likely that things went on very much as before except that particular attention was paid to finding a solution to the problems raised by incommensurability and the dialectical attacks of the Eleatics.

The solution eventually arrived at is usually attributed to Eudoxus (fourth century B.C.), a member of the Platonic academy in Athens. This was to make a distinction between numbers and magnitudes, and to provide definitions and theorems about ratios of magnitudes which replaced those expressed only in terms of numbers. The diagonal of a unit square was now regarded as a magnitude instead of as a length equal to the ratio of two numbers. It could be compared with and be in proportion to other magnitudes provided that these were of the same kind, that is, lengths could be compared with lengths, areas with areas, and volumes with volumes. Since lengths and areas were magnitudes of different kinds, they could not be compared. This was less flexible a situation than that of Babylonian mathematics. It avoided the problems raised by irrational numbers, but at the cost of separating number theory from geometry and introducing a dimensional rigidity which was to defer the remarriage of number theory and geometry for many centuries to come. Indeed, it was not until the seventeenth century A.D. that the breach began to be healed. The foundation of Greek mathematics was thus divided into two fundamental studies, *arithmetica* and *logistica*, the former being the approach from the point of view of the theory of numbers, the latter being the approach through magnitudes. According to late Greek commentators, *logistica* meant only the art of practical calculation, but these commentators failed to understand its true nature, and as a consequence have, until recently, misled scholars into underestimating its mathematical importance.

The introduction of the idea of ratios of magnitudes did not, however, avoid all the difficulties which arose from the dialetic of Zeno. If, for example, length is is a magnitude not made up of a string of ultimately indivisible points, then it must be infinitely divisible. We must be able to go on dividing it up into even smaller and smaller lengths without ever coming to an end of this process. Zeno had argued that if we want to reach a point B a given distance from a starting point A, we would first have to reach the point C, half-way between A and B. But before we can reach C, we must first reach the point D, half-way between A and C, and before reaching D, we must first reach E, half-way between A and D.

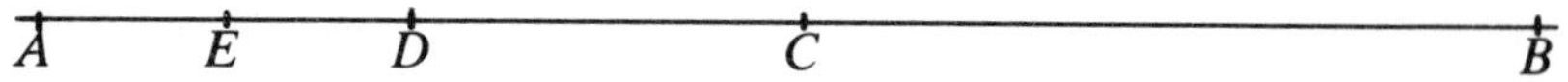

A E D C B

This argument can be pursued *ad infinitum*, hence we cannot ever get anywhere – motion is impossible! Zeno's paradox of the runner Achilles never being able to catch up with a tortoise if the latter is given a start is a similar argument.

The effect of these two paradoxes was to make Greek mathematicians highly suspicious of infinite processes and of infinity itself. They were prepared to accept that, given any number, however large, there would always be a larger number, and that given any line, however long, it could always be extended further. They were not prepared to accept the concept of an infinite collection of numbers nor that of a line of infinite magnitude, that is, whilst the concept of something being 'potentially' infinite was acceptable to them, they carefully avoided discussion of objects which were 'actually' infinite. The potentially infinite appears in two of the five postulates of Euclid. Postulate 2 reads:

> To produce a finite straight line continuously in a straight line.

Postulate 5, the famous 'parallel postulate' which was to give rise eventually to so many fruitful developments in geometry, includes the phrase: 'two straight lines, if produced indefinitely'. Nowhere, however, do we find reference to an infinite line as an entity in its own right.

The idea of infinity could not be kept out of mathematics indefinitely, however, and when Greek mathematical and philosophical writings became familiar in western Europe during the period of the Renaissance, arguments arose as to whether or not an infinite collection can exist. Difficulties came about because, as with the mathematical assumptions attacked by the Eleatics, the idea of such a collection gives rise to a number of paradoxes.

One of the first to draw attention to paradoxes associated with the idea of the infinity of the natural numbers was Galileo Galilei. In his *Discorsi* (1638), one of the characters, Salviati, raises the difficulty which arises when we attempt to count the number of perfect squares. If we count them in the usual way, that is, by one-to-one correspondence with the natural numbers, we have

$$\begin{array}{ccccccc} 1 & 4 & 9 & 16 & 25 & 36 & 49\ldots \\ 1 & 2 & 3 & 4 & 5 & 6 & 7\ldots \end{array}$$

Counting all the perfect squares seems to be using up all the natural numbers, so from this point of view the number of perfect squares is infinite and equal to the number of natural numbers. But, from another point of view, there must be more natural numbers than perfect squares because of all those that are not perfect squares:

$$2, 3, 5, 6, 7, 8, 10, 11, \ldots$$

This, and other similar difficulties of a geometric nature, led Galileo to reject infinite collections of objects as being 'not amenable to reason' since it was clear that relations such as 'greater than', 'equal to' and 'less than', as generally understood, could be applied only to finite collections. The crux of the problem lay in the acceptance of the dictum that 'a part cannot be equal to the whole'. Because it seemed that this did not hold good for infinite collections, such collections were rejected by philosophers and mathematicians alike.

It was not until the nineteenth century that mathematicians really came to grips with the problem of the actually infinite. Bernhard Bolzano, in a posthumous work, *Paradoxien des Unendlichen* (1851), defended the acceptance of actually infinite collections, and the fact that for such collections, as opposed to finite ones, the part can equal in number the whole. Georg Cantor and Richard Dedekind went further; they took this seemingly curious property and built it into the definition of an infinite collection. An infinite collection was thus a collection of objects equal in number to only a part of itself. The questions were still to be decided, however, as to whether there were the same number of rational numbers as there were natural numbers, and what exactly was the relationship of these numbers with irrational numbers, in particular with those irrational numbers which arose from the solution of polynomial equations with rational coefficients – equations such as

$$x^2 - 2 = 0$$

The answer to the former of these questions was provided by Cantor in 1874. We will not describe the argument in his 1874 paper showing that the rational numbers can be exactly counted by the natural numbers, since shortly after he provided a second argument which is less difficult to follow. This argument is based on the idea of a list – any collection which can be listed can be counted, provided that we have a way of ensuring that all its members are in the list. We can list all the rational numbers by first displaying them in an array. Along the first row, we list all those which have numerator 1, along the second row all those with numerator 2, along the third row all those with numerator 3, and so on:

$$\begin{array}{cccccc}
\frac{1}{1} & \frac{1}{2} & \frac{1}{3} & \frac{1}{4} & \frac{1}{5} & \ldots \\[2ex]
\frac{2}{1} & \frac{2}{2} & \frac{2}{3} & \frac{2}{4} & \frac{2}{5} & \ldots \\[2ex]
\frac{3}{1} & \frac{3}{2} & \frac{3}{3} & \frac{3}{4} & \frac{3}{5} & \ldots \\[2ex]
\frac{4}{1} & \frac{4}{2} & \frac{4}{3} & \frac{4}{4} & \frac{4}{5} & \ldots \\[2ex]
\frac{5}{1} & \frac{5}{2} & \frac{5}{3} & \frac{5}{4} & \frac{5}{5} & \ldots \\[2ex]
\ldots & \ldots & & \ldots & & \ldots
\end{array}$$

Now it is clear that we cannot count these row by row, as we exhaust all the natural numbers by one-to-one correspondence with the entries in the first row. It might seem therefore that, because counting each row exhausts the natural numbers and there are an infinity of rows, there must be more rational numbers than natural numbers. However, we can arrange to count in a different way, beginning with $\frac{1}{1}$, continuing with the rational numbers whose numerator and denominator add up to 3, then with those whose numerator and denominator add up to 4, and so on. This gives us a diagonal method of counting:

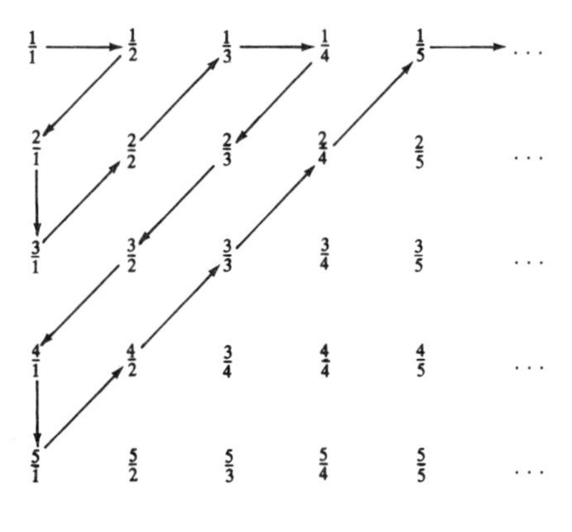

It is clear that this arrangement effectively enables us to list all the rational numbers. The fact that some of them occur more than once in different forms, such as $\frac{1}{1}, \frac{2}{2}, \frac{3}{3}, \ldots$ and $\frac{1}{2}, \frac{2}{4}, \frac{3}{6}, \ldots$, does not affect the force of the argument. Duplicates of this sort can be ignored once the corresponding form expressed in lowest terms has been taken into account. The rational numbers can be counted, that is, they can be put into one-to-one correspondence with the natural numbers. There are therefore the same number of rational numbers as there are natural numbers.

This seems at first sight to be an extraordinary fact, and not just because the array of rational numbers has an infinity of rows, each with an infinity of entries in it. It also seems extraordinary because the rational numbers possess the special property that between any two of them, however close together, we can always construct another. The natural numbers do not have this property; no natural number exists between 1 and 2, 2 and 3, or between any natural number n and its successor $n + 1$. There is, however, always the rational number $\frac{1}{2}(a + b)$ between any two rational numbers a and b. We can go on packing the rational interval between a and b with more and more rational numbers; this process never ends. At no point can we find a gap between two rational numbers which cannot be filled by other rational numbers. We therefore say that the rational numbers are 'dense' everywhere. Yet, Cantor proved that they can be counted by the natural numbers which do not share this property.

Cantor assigned the symbol $\aleph_0$ (aleph-null) to the infinity of the natural numbers. Thus we can speak of there being $\aleph_0$ natural numbers, 1, 2, 3, . . . , and $\aleph_0$ rational numbers p/q, where p and q are natural numbers. But, how many irrational numbers are there? – can these too be counted? If the rational numbers are dense everywhere, how is a number such as $\sqrt{2}$ to be included? Clearly, in some way, the density of the rational numbers between 1 and 2, does not exhaust all the numbers between 1 and 2 – we know that $\sqrt{2}$ is not accounted for.

We saw that it was proved by Gauss at the end of the eighteenth century that every polynomial equation has at least one root and that this is effectively the same as proving that every polynomial equation of degree n has n solutions. Solutions of equations had been known since the sixteenth century in Europe to include natural numbers, rational numbers, irrational numbers involving square, cube, fourth, etc. roots, and complex numbers. Excluding complex numbers, which clearly belong to a different category, the numbers which can be roots of equations extend those which we have so far stated to have been proved to be countable.

Cantor turned his attention to the collection of all algebraic numbers, that is, all numbers which can be the solution of polynomial equations,

$$a_n x^n + a_{n-1}x^{n-1} + \ldots + a_2x^2 + a_1x + a_0 = 0$$

with rational coefficients. These, too, he showed (in the same 1874 paper) to be countable, and, as these include all the irrational numbers involving square, cube, fourth, etc. roots, these must be countable also. We thus have three distinct collections of numbers, the natural numbers, the rational numbers, and the algebraic irrational numbers, each of which has $\aleph_0$ numbers.

Negative numbers had been rejected as solutions of problems in early times. They were eventually admitted in Hindu practical mathematics through monetary problems, since the idea of receiving and owing money was a simple and obvious one – a negative solution could be interpreted as a debt. Objection to negative numbers was, however, long-lived. As late as the early nineteenth century, we find works in which it is suggested that they do not have any real existence. The introduction of negative natural and rational numbers, does not affect the arguments about countability. We can list all the whole numbers, positive and negative, as follows:

$$1, -1, 2, -2, 3, -3, 4, -4, \ldots$$

and once they are listed, they can be counted. Thus, if the rational numbers which are ratios of natural numbers can be counted, so also can those which are ratios of positive or negative whole numbers. All these different collections of numbers have $\aleph_0$ members, and if we add together two collections, each with $\aleph_0$ members, we obtain a collection also with $\aleph_0$ members. This can be appreciated most easily by considering just the odd and even natural numbers. We can put the odd natural numbers into one-to-one correspondence with the natural numbers:

$$\begin{matrix} 1 & 3 & 5 & 7 & 9 & 11 & \ldots \\ 1 & 2 & 3 & 4 & 5 & 6 & \ldots \end{matrix}$$

We can do the same with the even numbers:

$$\begin{matrix} 2, & 4, & 6, & 8, & 10, & 12, & \ldots \\ 1, & 2, & 3, & 4, & 5, & 6, & \ldots \end{matrix}$$

Thus there are $\aleph_0$ odd numbers and $\aleph_0$ even numbers. Adding these two collections together we obtain the $\aleph_0$ natural numbers. Thus:

$$\aleph_0 + \aleph_0 = \aleph_0$$

The Real Numbers

Every schoolchild is taught that the ratio between the circumference and the diameter of a circle is a number which we call *pi* and denote by the Greek letter π, first used for this purpose by William Jones in 1706. This number, often approximated as $\frac{22}{7}$ or (less inaccurately) as $3{\cdot}14159\ldots$, is associated with the Greek problem of squaring the circle. This can take two forms. We may be required to calculate or construct the length of the side of a square whose area is equal to that of a circle of given diameter. Alternatively, we may be required to find, either by calculation or construction, the length of a straight line equal to that of the circumference of the circle. Both approaches effectively amount to the same problem. (We should remember here that only straight-edge and compass construction was permitted.)

Many attempts at squaring the circle have been made over the centuries. Indeed, despite the fact that it has been proved not to be possible, each new generation produces its crop of circle-squaring cranks.

Approximations to the ratio of the circumference to the diameter of a circle have been used from ancient times. In the Rhind papyrus, we find the calculation of the side of a square equal in area to a circle of given diameter. The result is equivalent to a value of π of $3\frac{13}{81}$. This is a much better approximation than the value 3 which is assumed by the writer of the Second Book of the Chronicles, where we find a molten sea 'round in compass' having a diameter of ten cubits and a circumference of thirty cubits. The best Babylonian approximation so far discovered is $3\frac{1}{8}$, and this approximation appears in early Hindu work. The familiar approximation, $3\frac{1}{7}$, derives from Archimedes, who showed that the circumference of a circle is less than $3\frac{1}{2}$ times but greater than $3\frac{10}{71}$ times the diameter. Archimedes' methods were taken up in western Europe by a number of mathematicians of the sixteenth and seventeenth centuries, one of the most notable approximations being that of Ludolph van Ceulen, who calculated the value of π correct to thirty-four places. Modern computers have enabled π to be calculated to more than a million places. We have thus been able to obtain ever better and better approximations to its true value. The question arises as to whether or not it is a rational number, and, if it is not rational, whether or not it is algebraic. The answers to these questions were not known until the eighteenth and nineteenth centuries respectively.

In 1761, Johann Lambert (famous for his contributions to problems associated with Euclid's fifth postulate – that on parallel lines) proved that π is irrational. This did not, of course, mean that squaring the circle was proved impossible, since algebraic irrational numbers, such as $\sqrt{2}$, are constructible with straight-edge and compass. Adrien-Marie Legendre conjectured that π could not be the root of any polynomial equation with rational coefficients, and, as a result, the existence of numbers which 'transcend algebra' gradually came to be discussed by mathematicians and to be known as 'transcendental' numbers. It was not until 1882, however, that π was finally proved to be one of these (by Ferdinand Lindemann) and the impossibility of squaring the circle, by systems of linkages let alone by straight-edge and compasses, long suspected by many mathematicians, was thus finally confirmed.

The number e, the base of natural logarithms was another obvious candidate for investigation. It was duly investigated and shown to be irrational by Euler, and in 1873 Charles Hermite, proved it to be transcendental, just nine years before Lindemann's corresponding proof for π.

We thus find that there are irrational numbers over and above those which are algebraic, and the question arises as to how many of these exist and in particular, if there is an infinity of them, can they be counted.

In 1844, Joseph Liouville proved that any number of the form

$$\frac{a_1}{10^1}+\frac{a_2}{10^{2.1}}+\frac{a_3}{10^{3.2.1}}+\frac{a_4}{10^{4.3.2.1}}+\ldots$$

where the numerators can be any natural number up to 9, is transcendental. Since there is an infinity of numerators each of which can take different values, this means that the number of transcendentals is infinite – strange though it may seem when we realise that only a comparatively few transcendental numbers have been discovered.

We can summarize this discussion by stating that we have four kinds of numbers (excluding complex numbers, *n* tuples, and arrays), and that each of these comprises an infinite collection. Three of these infinite collections, the whole numbers, the rational fractions, and the algebraic irrationals, are countable – their infinity is $\aleph_0$. We have yet to discuss the nature of the infinity of the transcendental numbers.

We call the collection of all four of these kinds of numbers the 'real numbers'. The countability or otherwise of the transcendental numbers is therefore a matter of the countability of the real numbers. If the transcendentals are countable then the real numbers must be countable since the sum of countable collections gives us a collection which is also countable.

The countability of the real numbers was investigated by Cantor and is discussed in the same 1874 paper in which the countability of the rational numbers and the algebraic numbers is proved. Cantor proved that the real numbers and hence the transcendentals are not countable, that is, cannot be put in one-to-one correspondence with the natural numbers. He gave more than one proof of this, the second and much the easier to understand appearing in 1890. It relies on the method of *reductio ad absurdum*.

Any decimal fraction which terminates after some specific number of places can immediately be seen to be a rational number. Thus $0{\cdot}5 = \frac{5}{10} = \frac{1}{2}$, $0{\cdot}44 = \frac{44}{100} = \frac{11}{25}$, and so on. In general, any decimal of the form $0{\cdot}a_1a_2a_3 \ldots a_n$ is just the rational number:

$$\frac{a_1a_2a_3 \ldots a_n}{10^n}$$

Any recurring decimal is also a rational number, though this is not quite so obvious. We are no doubt familiar with $0{\cdot}333\,33 \ldots = \frac{1}{3}$, $0{\cdot}111\,11 \ldots = \frac{1}{9}$, $0{\cdot}142\,857\,142\,857\,142\,857 \ldots = \frac{1}{7}$, and so on. Indeed, any rational fraction in its lowest terms, the basic factors of whose denominators are prime with respect to 10, will have a recurring decimal equivalent. (This is why the Babylonian base of 60 can be so useful – reciprocals of numbers having 3 as a basic factor in addition to just 2 and 5 terminate as sexagesimal fractions.) It is not difficult to see that every rational fraction will either terminate or have a recurring decimal expansion. When we divide the numerator by the denominator *d* in order to find a decimal expansion, we must invariably eventually be dividing into a series of zeros if the division does not terminate. There are only $d - 1$ possible remainders which we can have on division by *d* and so eventually we must repeat one of these. This must occur after at most $d - 1$ further divisions. Once we find a remainder being repeated, a recurring cycle must have been started. Consider a rational fraction with denominator 11, for example, say $\frac{8}{11}$. We have:

$$11\underline{|\,8{\cdot}800 \ldots}$$
$$\;\;\;\;0{\cdot}727 \ldots$$

The remainder 7 is repeated after only two further divisions, and we know then that the decimal expansion is $0{\cdot}727\,272 \ldots$. Here, the cycle consists of just two digits, 7 and 2. In the case of the denominator 7, we have the longest possible cycle of $n - 1$, a cycle of $7 - 1 = 6$ digits.

We still need to show that every recurring decimal between 0 and 1 is a rational number. To see that this is so, we will first look at a specific example, the decimal fraction

$$0{\cdot}123412341234\ldots$$

We note that there is a recurring cycle of four digits. If therefore we multiply by 10^4 and subtract, the recurring cycle will disappear:

$$\begin{array}{r} 1234{\cdot}1234\ 1234\ 1234\ldots \\ 0{\cdot}1234\ 1234\ 1234\ldots \\ \hline 1234\phantom{{\cdot}1234\ 1234\ 1234\ldots} \end{array}$$

So, if we denote the original number by n, we now have

$$10^4n - n = 9999n$$

The recurring decimal is therefore equivalent to the rational number $\frac{1234}{9999}$.

This method is applicable to any recurring decimal whether or not it includes non-recurring digits in the earlier decimal places. Thus, for

$$0{\cdot}212\ 438\ 438\ 438\ldots,$$

we multiply by 10^6 and by 10^3 and subtract:

$$\begin{array}{r} 212\ 438{\cdot}438\ 438\ldots \\ 212{\cdot}438\ 438\ldots \\ \hline 212\ 226\phantom{{\cdot}438\ 438\ldots} \end{array}$$

We now have:

$$10^6n - 10^3n = 999\ 000n$$

The recurring decimal is therefore the rational number $\frac{212226}{999000} = \frac{35371}{166500}$.

In general, the recurring decimal

$$0{\cdot}a_1a_2\ldots a_jr_1r_2\ldots r_kr_1r_2\ldots r_kr_1r_2\ldots$$

where there are j initial non-recurring digits followed by k recurring units, can always be expressed as a rational number by multiplying by 10^{j+k} and 10^k, subtracting, and dividing by $10^{j+k} - 10^k$. Since we have an algorithm for the general case, we know that every recurring decimal is a rational number. It would be nice and tidy to be able to add 'and vice versa'. We can do this in two ways – either by appending zeros indefinitely once a decimal expansion terminates, so that, for example, 0·5 becomes 0·50000 . . . , or by subtracting 1 from the last decimal digit and appending nines indefinitely, so that 0·5 becomes 0·4999 . . . . This is perfectly legitimate since the algorithm above shows us that 0·999 . . . = 1; hence 0·09999 . . . = 0·1 and 0·4999 . . . = 0·5.

Since we have now accounted for all the rational numbers by showing that these must be terminating or recurring decimals and vice-versa, we are left with the irrational numbers (including the transcendentals) and the infinite decimal expansions which are not recurring. These must therefore correspond. We can now return to Cantor's proof that the real numbers cannot be counted.

We again consider just the real numbers between 0 and 1. Each of these can be written as an infinite decimal as we have seen. If these real numbers are countable we must be able to arrange them in a list in which each appears in its proper place just once:

$$\begin{array}{l} 0{\cdot}a_{11}a_{12}a_{13}a_{14}a_{15}\ldots \\ 0{\cdot}a_{21}a_{22}a_{23}a_{24}a_{25}\ldots \\ 0{\cdot}a_{31}a_{32}a_{33}a_{34}a_{35}\ldots \\ 0{\cdot}a_{41}a_{42}a_{43}a_{44}a_{45}\ldots \\ \quad\ldots\ \ldots\ \ldots\ \ldots \end{array}$$

Let us suppose that there is a way in which such a list can be determined, that is, we suppose that the real numbers between 0 and 1 are countable. We now construct a real number,

$$0{\cdot}b_1b_2b_3b_4b_5\ldots$$

such that the digit b_1 differs from a_{11}, b_2 differs from a_{22}, b_3 differs from a_{33}, and so on. This number, also lying between 0 and 1, differs from every number in our list by at least one digit. It is therefore not in the list. This contradicts our assumption that we can list all the real numbers between 0 and 1 and hence also that they are countable.

The real numbers are uncountable – their infinity is not the same as that of the natural numbers, the rationals and the algebraic irrationals. And, as this must be due to the inclusion of the transcendentals, it is these which are uncountable. Cantor denoted the infinity of the real numbers by c, the first letter of 'continuum'. Clearly we can say that c is greater than $\aleph_0$ provided that we take care in defining precisely what we mean by 'greater than' in the context of infinite numbers. We do this by noting that, whereas $\aleph_0$ numbers can be put in one-to-one correspondence with only a part of the c real numbers, there is no way in which c numbers can be put in one-to-one correspondence with a part of $\aleph_0$ numbers. This applies, of course, to any collections of $\aleph_0$ and c objects – it is a fundamental truth of what we call 'set theory', the general theory which derives from Cantor's work, and is not confined to collections of numbers.

Numbers such as $\aleph_0$ and c are called 'transfinite' – they are 'beyond the finite' – and there are infinitely many of them. This was proved by Cantor by showing that given any cardinal number, finite or transfinite, it is always possible to construct one that is greater.

Cantor's work on infinite sets thus leads us to the situation that, although the rational numbers are dense – however small an interval we choose, it will always contain an infinity of them – this denseness is nevertheless full of 'gaps', which are in turn filled by the transcendental numbers. Even harder to appreciate, perhaps, is the fact that there are more transcendental numbers filling these gaps than there are rational numbers. It is not surprising that Cantor's work was not immediately accepted by his fellow mathematicians. Their suspicions of it were heightened when it was discovered that, although set theory did solve some of the difficulties which had previously been associated with the real numbers, it raised other difficulties and paradoxes of its own. It was described by some of his contemporaries as 'a disease', 'an abyss of transcendentals', 'repugnant to common sense', and so on. These attacks depressed Cantor and led to a series of nervous breakdowns. He died in 1918 in a mental institution in Halle. It was not until the present century that the genius of his work was properly recognized.

We have seen, then, that it was shown that the infinity of the real numbers is not the same as that of the rational numbers. There was, however, another important question to be answered, namely 'what is an irrational number?' There was no difficulty in explaining rational numbers – they were just ratios of natural numbers. The understanding of what numbers are could not be complete unless some way was discovered to explain irrationals in terms of rational numbers. There would then be a satisfactory sequence of explanation. Starting with the natural numbers, the whole numbers (positive and negative, and zero) were explained as pairs of natural numbers which were subtracted: -3, for example, arose from a natural number pair such as 2,5 through subtraction. Rational numbers were then explained as pairs of whole numbers which were divided: $\frac{2}{3}$, for example, arose from a whole number pair such as 2,3 through division. There the sequence of explanations stopped, though it could be taken up again by explaining complex numbers as pairs of real numbers a,b.

The question of the explanation of irrational numbers in terms of the rationals was investigated by both Cantor and Dedekind. Cantor's approach was to consider sequences of rational numbers. This was by no means an entirely new idea. Such sequences had been investigated earlier in the nineteenth century by Gauss, Cauchy, Meray and others. Dedekind's approach was from the point of view that each real number n divided the collection of all real numbers and hence also the collection of all rational numbers into two parts – numbers less than or equal to n and numbers greater than n. He called such a division of the rational numbers a 'cut'.

If we make a cut in the rational numbers by selecting a real number n, two possibilities arise. If n is itself a rational number, then we can assign it to that that part of the total collection of rationals which consists of all those less than or equal to n, and this part will then have n as its greatest member. So, if we take n to be $\frac{3}{5}$, this divides the collection of all the rationals into those less than or equal to $\frac{3}{5}$ and those greater than $\frac{3}{5}$, and the former will have $\frac{3}{5}$ as its greatest member. If, however, n is irrational, it will not belong to the rationals which are less than or equal to it, and consequently no greatest member will be defined. Thus, if we take n to be $\sqrt{2}$, we still divide the rationals into two, but $\sqrt{2}$ does not belong to the lower part (nor to the upper part), and no greatest member of that part is defined.

Dedekind's approach enables us to define an irrational number as a special kind of cut in the rational numbers. To show that this is an acceptable definition requires that we prove also that such cuts behave in every way like real numbers – that all the operations performable on real numbers can be performed on the cuts with the same results. We shall not go into this further. The aim has been to show that there are ways in which the apparent gap in the explanation sequence leading from the natural numbers to the complex numbers can be closed.

Dedekind's approach to defining the irrational numbers shows some similarities with the way in which the Greeks, and Eudoxus in particular, evaded the problems associated with the discovery of incommensurably. In Book V of Euclid's *Elements*, we find the following definition of the ratio of two magnitudes:

> Magnitudes are said to be in the same ratio, the first to the second and the third to the fourth, when, if any equimultiples whatever be taken of the first and third, and any equimultiples whatever of the second and fourth, the former equimultiples alike exceed, are alike equal to, or are alike less than the latter equimultiples taken in corresponding order.

This somewhat tortuous explanation, attributed to Eudoxus, effectively states that two ratios of magnitudes, $a:b$ and $c:d$ are equal only if, given natural numbers m and n,

$$mc < nd \quad \text{whenever} \quad ma < nb$$
$$mc = nd \quad \text{if} \quad ma = nb$$

and

$$mc > nd \quad \text{whenever} \quad ma > nb$$

This does have certain similarities with Dedekind's definition of real numbers since it separates the collection of all rational numbers into two parts according to whether ma is less than or equal to nb, or greater than nb. The interpretation of these similarities has sometimes been overplayed. It is not correct to claim that Eudoxus, by avoiding the problem of irrational numbers, anticipated Dedekind's definition by over 2000 years. This is to read ancient history with hindsight provided by the mathematics of the nineteenth century, and thus to see connections which are mathematically but not historically justifiable.

The Status of Zero

We saw that the demand for a symbol to represent an empty place arises only in place-value systems of numerals. A zero symbol duly made its appearance in Babylonian mathematics, though this was relatively late and it was used only in a limited way. With numerals such as those of Egypt and Greece no zero was necessary. The Greeks were nevertheless aware of the concept of zero, though it had no place in their theory of numbers. The reason for its rejection as a number was clearly stated by Aristotle to be that it could not be used in forming ratios. Even after a zero symbol had been incorporated into the Hindu written system, we find references to nine numerals rather than to ten. Aristotle had indeed pointed to the difficulty inherent in accepting zero as a number – we cannot divide by it. Nevertheless, we find an attempt to explain the division of zero by itself in the writings of Brahmagupta, and Bhâskara uses division by zero as a way of defining infinity.

The status of zero as a number remained unsettled long after the introduction of the Hindu–Arabic numerals into western Europe. This is indicated by the different names by which it was known, and the fact that one of these, *cifra*, was used as a secret word by esoteric societies. The ordinary people regarded zero as being endowed with magical properties. After all, although it was 'nothing', readily understood when it was added or subtracted, when it stood next to a number it increased its value tenfold. Division by zero was somehow dimly understood as leading to the infinite, and the infinite was God.

Even mathematicians shared the common suspicion of zero. Although they might list it as one of the ten numerals of the Hindu–Arabic system, they usually added remarks which indicate that it had a status different from all other numbers. Indeed, it is doubtful if it was thought of as being a 'number'; it was, rather, the denial of number. At the same time, there was no real difficulty experienced in using it in arithmetic operations. It was not its use but its status which was called into question. The fact that it is not used in counting no doubt contributed considerably to all this uncertainty.

To some extent, zero as a concept, retains a special status today. We are accustomed to accepting it as a member of the sequence of integers

$$\ldots -3 \; -2 \; -1 \quad 0 \quad 1 \quad 2 \quad 3 \quad \ldots$$

and as the value of the variables at the intersection of the axes of a graph. It plays an important role as the identity element for addition (and subtraction) equivalent to the role played by unity for multiplication (and division). This is simply to express in formal language that zero added to (or subtracted from) any number leaves that number unchanged in the same way that any number multiplied or divided by one remains unchanged. Zero has therefore come to have a status as a number which in a certain particular way corresponds to that of the number one. This status becomes changed, however, as soon as we introduce division. Zero has then to be excluded – it is no longer acceptable as equal in status to other numbers.

Today we are still not able to answer the question 'is zero a number?' without qualification. For some purposes we can regard it as just one number amongst others, for other purposes we cannot – and it has to be excluded. It therefore continues to have something of a unique status of its own. Indeed, Aristotle was entirely correct in rejecting it from the numbers discussed and used in Greek mathematics, and his reason was precisely the reason which continues to give it its special status today.

Complex Numbers

We saw that complex numbers arose out of attempts to find all the solutions of cubic equations. They were by no means readily accepted by mathematicians, however. Indeed, even negative solutions to equations were still being rejected by a few mathematicians as late as the early nineteenth century. Complex numbers were described as 'useless' or 'non-existent' long after it had been accepted that certain equations have complex roots. They were thus usually referred to as 'imaginary' numbers, a name which is still occasionally used today.

During the eighteenth century complex numbers were introduced into a number of areas of mathematics, particularly those associated with the calculus. They were thus used in many mathematical calculations although, it is fair to say, they were by no means well understood. Three mathematicians in particular were responsible for removing the cloak of mystery with which complex numbers were surrounded; these were Caspar Wessel, Jean-Robert Argand, and Gauss. Wessel, a Norwegian surveyor and largely self-taught mathematician, proposed in 1799 that complex numbers should be represented by points in a plane by taking the number 1 as the unit on a horizontal axis and $\sqrt{-1}$ as the unit on the corresponding vertical axis. Any complex number $a + ib$ is thus represented by the point P, b units above the point a on the horizontal axis. The addition of a and ib is thus interpreted vectorially. We move a distance in one direction and then a further distance in a different direction and our final position from our starting point is obtained by the vector sum of these two distances – direction is taken into account. Wessel was then able to go on to interpret all the basic operations on complex numbers in terms of lines in the plane of a and ib.

Argand's approach, published in 1806, achieved much the same result.

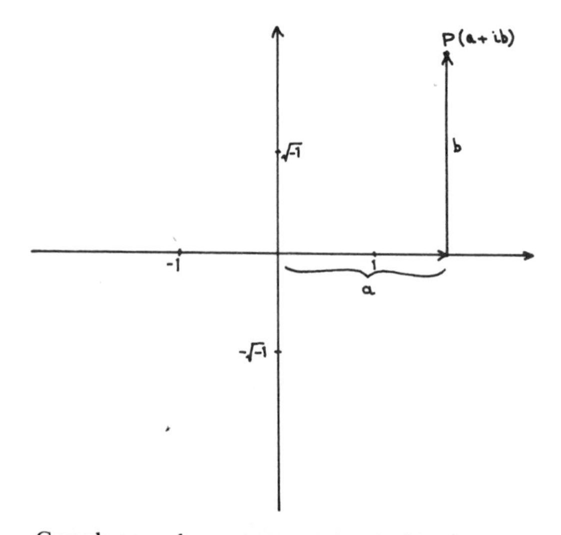

Complex number representation in form $a + ib$

However, he began by considering operations on a line segment of length a lying along a single (horizontal) axis.

0 ⎯⎯⎯⎯⎯⎯⎯⎯→
a

Reversing the direction of a about the origin 0 can be interpreted as multiplication by -1. This establishes the points a and $-a$.

$-a$ 0 a

Thus, multiplication by -1 is equivalent to rotation through 180 degrees. Multiplication by -1 a second time, corresponding to a further rotation by 180 degrees, just restores a. Successive multiplication of a by $i = \sqrt{-1}$, however, gives us four numbers: $a, ia, -a$ and $-ia$. If we now regard multiplication by i as counter-clockwise rotation by 90°, we have a situation which is consistent with that when we successively multiply by -1, and represents ia as a point a units vertically above the origin and $-ia$ as a point a units vertically below the origin. This suggests the alternative way of writing down these four points as $a\angle 0$, $a\angle 90$, $a\angle 180$, and $a\angle 270$ (or $a\angle -90$). Here, we have the points expressed in terms of distance from the origin and a counter-clockwise angle of rotation from the positive horizontal axis. This is easily extended to representing any complex number $a + ib$ in the form $r\angle\theta$ where r is the distance of the point P from the origin O and θ is the angle of rotation from the positive horizontal axis. The relationships of r and θ to a and b are given by

$$r = +\sqrt{a^2 + b^2}$$

$$\tan\theta = \frac{b}{a}$$

as indicated in the diagram.

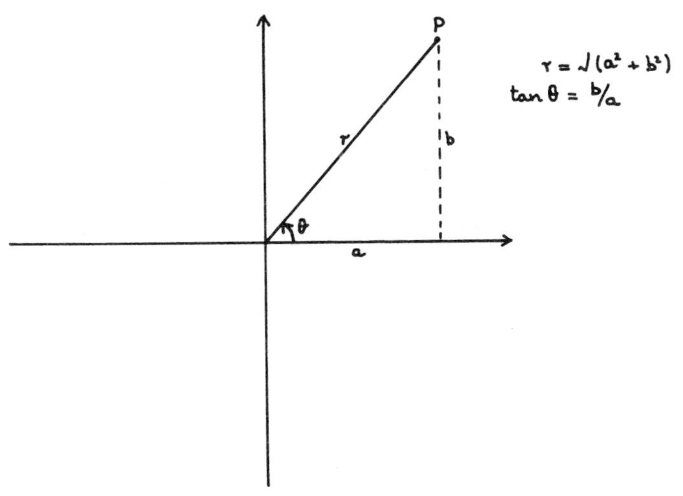

Complex number representation in form $r\angle\theta$

Operations on complex numbers can now be carried out on whichever form of representation is convenient. Addition and subtraction are easier to carry out in the $a + ib$ form. We have

$$(a + ib) \pm (c + id) = (a + c) \pm i(b + d)$$

that is, we add the real and complex parts separately. Multiplication and division are easier to carry out in the $r \underline{/\theta}$ form. We have

$$r \underline{/\theta} \times s \underline{/\phi} = rs \underline{/\theta + \phi}$$

and

$$r \underline{/\theta} \div s \underline{/\phi} = \frac{r}{s} \underline{/\theta - \phi}$$

that is, we multiply (or divide) the lengths and add (or subtract) the angles.

However, complex numbers can be regarded as pairs of real numbers, that is, we can regard $a + ib$ as the pair of numbers a,b – not, we should note, the same pair as b,a. All the operations on complex numbers can now be redefined simply in terms of number-pairs and the imaginary i, representing $\sqrt{-1}$, can be dropped altogether. This approach was due to Gauss. It was later taken up by Hamilton in his search for triples of numbers which would correspond to points in space in the same way that pairs of numbers correspond to points in the plane, and which could be manipulated in ways corresponding to the operations on complex numbers. In fact, after many years studying this problem, Hamilton eventually discovered that this was not possible. He had to extend from triples of numbers a,b,c to quadruples of numbers a,b,c,d and also to give up the commutative property of multiplication. This meant that, for two quadruples A and B, it was not generally true that $AB = BA$. The behaviour of these 'quaternions' thus represents a fundamental departure from the laws of arithmetic governing both the real and complex numbers. Although made up from real numbers, they did not behave entirely like real numbers.

Defining the Natural Numbers

In all our discussions, we have taken the natural numbers for granted. The other kinds of numbers discussed have been shown to be explainable in the last resort in terms of the natural numbers. Until the nineteenth century, the natural numbers were always taken as in some sense 'given'. Many mathematicians still take this view today. They would more or less echo the words of one of Cantor's greatest critics, Kronecker, who is on record as stating: 'God created the integers, all the rest is the work of man'. Quite apart from accepting negative as well as positive whole numbers as being given, this statement raises a number of theological and philosophical problems which we cannot explore here, though the question of whether man 'discovers' or 'creates' new mathematics and science is no doubt one that will be debated to the end of time.

It was not until late in the nineteenth century that formal definitions of the natural numbers began to be sought. In 1884, Gottlob Frege, published his *Foundations of Arithmetic*, in which an attempt was made to define them from the standpoint of counting. Counting essentially involves the principle of one-to-one correspondence. Any two collections which can be put in one-to-one correspondence have the same cardinal number. Frege said that the collection of all collections, that is, the set of all sets which can be put into one-to-one correspondence with each other defines a cardinal (i.e. natural) number. Thus the number five is defined by the set of all collections of five objects. Frege's original definition of numbers was, however, not free from difficulties. In particular, paradoxes arose as a result of accepting the idea of a set of all sets, which lay at the heart of his definition.

Five years after Frege's definition had appeared, another approach was published by Giuseppe Peano. This depended on the basic idea of the succession of the natural numbers one after the other – an idea which is a prerequisite to counting. We can start with unity and define it to be the first natural number. We can then state that every natural number has a successor and that two natural numbers with the same successor are equal. Finally, we have to state:

> If a set of natural numbers includes unity and the successor of every number which is in it, then it includes every natural number.

We can thus define every natural number in terms of unity. If we denote the successor of a natural number n by $S(n)$, we can define two as $S(1)$, three as $S(2) = SS(1)$, and in general, any natural number n as

$$\underbrace{SSSS\ldots S}_{n-1}(1)$$

where there are $n - 1$ Ss and brackets have been omitted to avoid complication. From this argument it would seem to follow that, provided that we accept the existence of unity, all the natural numbers exist and can be derived from unity using the concept of succession.

This definition of the natural numbers underlies a method of proof, peculiar to mathematics, known as 'proof by mathematical induction'. Despite its name, this is a deductive method of proof, inductive methods being foreign to mathematics. The method involves three stages. First, we have to show that if we assume that a statement is true for an unspecified variable n, it then follows that the same statement must also be true for $n + 1$. Next, we have to show that the statement is in fact true when $n = 1$. What we have done so far, therefore, is to show that the statement is true for 1 and for every successor of 1. The Peano definition of the natural numbers now allows us to argue that the statement must therefore be true for every natural number.

The Peano approach represents a considerable step forward in the understanding of numbers. There was, however, yet a further stage of abstraction possible.

Cantor's work had suggested that there might be a purely set-theoretic definition of the natural numbers. This idea was of special appeal to those who saw set theory as the basis of both mathematics and logic. There were also those who objected to the definitions of Frege and Peano on the grounds that they began with one or more of the very objects which they were supposed to be defining. Defining the natural numbers on the basis of unity, for example, accepts unity as a natural number *ab initio* and before what we mean by natural number has been defined. There was thus seen to be an unacceptable circularity about such approaches. The way to avoid this was not to start with anything, that is, to start with nothing.

The most basic set of all is the empty set, the set with no members. This set is usually denoted by the symbol ø (a Scandinavian not a Greek letter). We will define zero to be this set. We now define the natural number one as the set with just the member ø. We then define two to be the set with just the members ø and the set with the one member ø, three to be the set with just the members ø, the set with ø as its only member, and the set with both these as members, and so on. Using curly brackets to enclose the members of a set, we thus have:

$$0 = \emptyset$$
$$1 = \{\emptyset\}$$
$$2 = \{\emptyset, \{\emptyset\}\}$$
$$3 = \{\emptyset, \{\emptyset\}, \{\emptyset, \{\emptyset\}\}\}$$

and so on. The natural numbers are thus definable, as it were, from nothing. This represents the peak of abstraction so far achieved in their definition. Strictly, this twentieth-century definition, due to John von Neumann, defines ordinal numbers. The cardinal numbers can, however, be defined from them. There is no circularity in this approach since we begin not with zero or any natural number, but with the concept of the empty set.

Numbers and Religious Belief

It has been remarked elsewhere that the creation of the natural numbers from the empty set, that is, from nothing is a theological rather than a logical approach, matching the creation of the universe out of nothing described in the first chapter of Genesis. Be that as it may, it is certainly true that throughout man's history there has been a close connection between the natural numbers and the activities of the divine Creator. The natural numbers have always been seen by some at least in every civilization to be symbolic of deep esoteric religious beliefs. The Babylonians had a hierarchy of sixty gods each of which was associated with one of the first sixty natural numbers. The place of any particular God in this heavenly hierarchy was indicated by his number. In ancient India, we find religious significance assigned to each of the first 101 numbers, and in the Mayan civilization the first thirteen numbers all represented gods.

There is something of a parallel in other sciences. Astronomy grew from astrology and chemistry from alchemy. It is therefore not surprising that number theory, first developed by the Pythagoreans, was associated with their religious beliefs and practices.

Numbers were central to the Pythagorean view of creation. One was the basic unit from which all others were created – it therefore was symbolic of the divinity, of reason, and of all that was eternal and unchanging. Even numbers were feminine and symbolized things essentially belonging to the earth. Odd numbers were masculine and symbolized things belonging to the heavens. Four, the first square number, represented justice – perhaps it is from this that we get our expression 'a square deal'! Five represented marriage, the union between the first female number, two, and the first male number, three.

The Pythagoreans reverenced the *tetraktys*, a triangular array representing the first four numbers and adding up to ten, the number of perfection. It also symbolized the harmony of the universe and the unity of the four basic elements – fire, water, earth and air. New members of the brotherhood had to swear a secret oath by the *tetraktys*, which also acted as their password. Prayers were offered to it. It was invoked as the 'holy, holy *tetraktys*, that containest the root and source of the eternally flowing creation'. In order to fit in with this number of perfection, the Pythagoreans added to the known seven 'planets' (which included the sun and the moon) a complete sphere of fixed stars, the earth itself, and an invisible counter-earth, thus making the heavenly bodies up to ten. Many of the Pythagorean beliefs about particular numbers have survived to the present day amongst occultists.

Such beliefs were not confined to the Greeks. We find similar beliefs in ancient China and amongst the Jews. For the Chinese, odd numbers represented life and heat, even numbers death and cold. Thus odd numbers were represented by white dots and even numbers by black dots. The *lo-shu*, the oldest example of a magic square, included all the natural numbers up to nine represented in this way, and was thought to have magic properties. For the Jews, seven was of special importance. It occurs many many times in the Old Testament. In the Book of Joshua, we read that Joshua was instructed by God as follows:

> And seven priests shall bear before the ark seven trumpets of ram's horns: and the seventh day ye shall compass the city seven times.

Again, we find that Jacob bows seven times to the ground before Esau, that houses are ritually cleansed by being sprinkled seven times, that the Shunammite child raised by Elisha 'sneezed seven times', that Naaman has to wash in the Jordan seven times to be cured of his leprosy. There are literally scores of such references to seven.

The Jewish predilection for the number seven passed into Christian writings. Seven devils were cast out of Mary Magdalene; seven loaves fed the four thousand; seven deacons were chosen to have oversight of the financial and charitable activities of the early Church; and in the Apocalypse we have the seven churches, the seven golden candlesticks, the seven angels, the seven vials, the seven kings, and so on. We could continue with the seven deadly sins, the seven joys of the Virgin, the seven spirits of God, and many more such examples. We do not know just how seven came to have such mystical significance. A suggested reason is put forward in the writings of the Jewish philosopher Philo (first century A.D.), who states that if we take just the first ten numbers, only the number seven is neither 'produced' (i.e. has factors) nor 'produces' (i.e. is a factor of another number). In addition, he points out that when seven is multiplied by four, it produces twenty-eight, a perfect number, that is, a number equal to the sum of its divisors.

The other perfect number known in ancient times, six, was also of special significance for both Jewish and Christian writers. We read of the six days of creation, the six lambs without blemish, Goliath's height as six cubits, the six boards of the tabernacle, the six steps of King Solomon's throne, the six measures of barley, and so on. Along with seven, the number six also passed into the mystical deliberations of Christian writers. Saint Augustine even went so far as to suggest that six as a perfect number has an existence independent of creation. He wrote:

> God created all things in six days because this number is perfect, and it would have been perfect even if the work of the six days did not exist.

The numbers two and forty also had a special significance carried over from Old Testament to New Testament writings. We have the two animals of each kind of preserved in the ark, the two portions of land assigned by God to Joseph, the two anointed ones, the sending forth of the disciples two by two, and the two witnesses. There is also the forty days and nights spent by Moses on Mount Sinai, the forty years spent by the Children of Israel in the Wilderness, and the forty days and nights of Jesus' fasting and temptation in the wilderness, and many other examples far too numerous to be quoted here.

In the theological writings of the Eastern Church, we also find many references to numbers in a mystical context. The authors of the *Philokalia* state that they have divided the famous discourse on prayer into 153 texts because 153 is both a triangular and a hexagonal number, the triangle being representative of the Trinity and the hexagon of the creation of the world in six days. They then go on to explain the mystical significance of 5, 25, 28, 53 and 100.

We have seen that, as a result of the use of letters of the alphabet as numerals, it became fashionable to associate words directly with numbers. This type of gematria, possible with both Hebrew and Greek, reinforced the determination of Biblical interpreters to preserve the mystical significance of both overt and covert numbers to be found in the Scriptures. Renaissance enlightenment, far from relegating such interpretations to pseudo-theology, reinforced the view that numbers in some way really do hold a key to the understanding of the Bible. Poets and secular writers also made considerable use of number symbolism, but it was primarily theological writers who rivalled each other in finding more and more evidence of the significance of Biblical numbers. This continues today on the fringes of Christian orthodoxy, but in the nineteenth century it was still a respectable part of scriptural exegesis.

One particularly remarkable work appeared as recently as 1917. The Reverend A. A. Bramley-Moore, in a work entitled *The Significance of Numbers as used in the Bible*, devotes a chapter to each of the numbers 1 to 15, 17, 19, 20, 30, 31, 37, 40, 42, 43, 50, 70, 153, 200 and 666, (the number of the beast). Drawing on earlier works by Bullinger and Panin, he virtually exhausts the possibilities of numerical scriptural exegesis. One, he says, denotes the sovereignty of God, and he quotes the obvious passages which underline the monotheism of the Jewish religion. This is all in accord with the Pythagoreans. Two, however, is symbolic of incarnation. This is not unreasonable in view of the fact that Jesus is for Christians both human and divine. However, gematria is brought in to show that the various names for Christ in the New Testament, and also the names of those in the Old Testament who are said to typify Christ, have numerical values in which two is the main factor, that is, is repeated more than any other. Three is, of course, symbolic of the Christian Trinity. But the author goes on to make much of the fact that three languages, Latin, Greek and Hebrew, were used in the inscription on the Cross. These are said to be symbolic of the leaders of the political, intellectual and religious world, all of whom, says the Author, have been at war with Christ ever since. It is with thirteen, however, that Bramley-Moore excels himself. This is said to signify rebellion, apostacy and sin. He finds that many of the enemies of God have had names related directly to thirteen. He quotes:

Bigthan = 455 = 13 × 35
Haman = 117 = 13 × 9
Zeresh = 507 = 13 × 13 × 3
Elymas = 676 = 13 × 13 × 4
Hermogenes = 481 = 13 × 37
Philetus = 1181 = 13 × 86
θηρ (a wild beast) = 117 = 13 × 9
θηριον (beast) = 247 = 13 × 19
'He had two horns' = 1521 = 13 × 13 × 9

He is not averse to stretching things a little. Simon Magus has to be 'with article' in order that he may equal 1170 = 13 × 90. Sometimes whole sentences or parts of sentences are used. Thus 'The man who took not God for his strength' is equal to 2197 = 13 × 13 × 13 as are also 'The chief of transgressors' and Jesus' first recorded words 'How is it that ye sought me, wist ye not'.

Yet this curious work is by no means entirely without scholarship. Quite apart from an extensive knowledge of the Bible in Hebrew and Greek, the Author is clearly familiar with the ordinal and cardinal forms of alphabetical numerals, and with the typological interpretation of Old Testament figures – the last, no doubt, on account of his upbringing in a Catholic Apostolic household. This particular Church developed Old Testament typology more exhaustively than any other ecclesiastical body.

The aversion to the number thirteen is probably as widespread today as it has ever been. Many hotels have no room 13, many hospitals no bed 13, many airports no departure gate 13, and so on. Often, the number 12a is substituted for 13. When, on 13 April 1970, the Apollo 13 mission was reported to be having problems, it was pointed out that its launching time had been 1313 hours, its launching pad $39 = 3 \times 13$, and sleeping periods for the astronauts were timed for 13 minutes past the hour – who could then doubt but that the mission would prove to be a disaster!

All sorts of explanations have been suggested for the universal aversion to 13 as being unlucky, evil, or symbolic of death. Psychologists claim that it is a matter of conditioning – we are brought up in an environment in which 13 has certain connotations and we are conditioned by this. This may well account for our being suspicious of 13 today, but it does not suggest how these suspicions first began.

The Reverend Bramley-Moore provides us with all kinds of possible explanations, and he may not be so wide of the mark after all. The number thirteen occurs five times in the Old Testament and, in addition, there are three references to the 'thirteenth year', and six to the 'thirteenth day' in the Book of Esther. Perhaps especially significant is the absence of any reference to the 'thirteenth month', possibly either on account of its significance in relation to the female menstrual cycle or because of association with pagan religion through the lunar year. The writer of Esther would have been familiar with the thirteen lunar months since he wrote shortly after return from the captivity in Babylon. To mention a thirteenth month directly might have been too overt a reference to the captivity experiences.

It certainly does seem that from the time of the Book of Esther onwards (from the second century B.C.) the number thirteen became gradually accepted as synonymous with doom and evil. Earlier references to thirteen in the Bible do not have these associations. Ishmael, Abraham's son, is said to have been circumcised in his thirteenth year, thirteen bullocks formed part of the prescribed meat offering to be made to God, Solomon's house was built in thirteen years, the 'sons and brethren of Hosah' numbered thirteen in all, and the length of the gate of the city of Ezekiel's vision was thirteen cubits. There was a rebellion against the King of Elam by certain subject kings after thirteen years, but this was hardly a disaster, and the remaining two Old Testament references to thirteen relate to the word of the Lord coming to Jeremiah.

This provides only a possible explanation of the origin of our dislike of thirteen. It is also possible that in the future biologists will discover that the number is connected in some way with biological rhythms of the human body which spark off in our subconscious a fear related to the concept of thirteen. Whatever the true explanation, thirteen does have effects on present-day society. It has been estimated that it costs the American economy some billion dollars each year in loss of business on the twelve days of the year which fall on the thirteenth of the month. Corresponding figures are not available for Britain, but it is not unlikely that the cost to the British economy is also significant. We have little reason to doubt that numbers, for all their mathematical development, still play an important role both in our formal religious beliefs and also in the many superstitions which flourish along with those beliefs, even though this has been described as the century of progressive atheism.

Chapter Nine

Teaching and Learning Numbers

> The life so short, the craft so
> long to learn.
>
> (Hippocrates)

> Delightful task! to rear the tender thought,
> To teach the young idea how to shoot.
>
> (James Thomson)

We do not have an indefinite amount of time available to us in this life, though we often behave as if we do. Time, like all other resources at our disposal, is limited and we should make good use of it. This is not to suggest that we should spend it exclusively on things which can immediately be seen to be productive. It does suggest, however, that it is preferable that our recreational and leisure activities should be creative in some sense at least.

This principle applies to our study of history, whether it be political history, church history, economic history, history of science, or history of mathematics. All history involves the history of ideas, and this is especially true of the history of mathematics. It is, of course, fascinating to study ideas just for their own sake, but, in the context of our finite life-span, we ought to be alert and responsive to those aspects of history which can suggest to us ways of enriching our experience of life. No matter what kind of history we study, we ought to find it rich in lessons applicable to the present. Study of the history of mathematics is especially rewarding in this way. It gives us a much deeper understanding of mathematics itself than is possible from a study only of the mathematics conventionally taught in schools and universities, since it enables us to appreciate how today's mathematics has evolved. Indeed, it can draw our attention to the very processes of mathematical evolution and suggest to us how these processes are related to their general social and cultural environment. It can also tell us how fashions in the teaching of mathematics change over the centuries, and it suggests that both the methods used and the content of what is taught should be highly flexible – suited not only to the particular environment in which teaching takes place but also tailored to fit the abilities and the disabilities of those being taught. History of mathematics provides a rich source of alternatives for those teachers who are genuinely concerned about the problem of numeracy today, and are concerned also when individuals in their charge appear to have difficulties with the mathematical ideas and techniques in the particular forms in which they are being presented to them.

It is not possible here to attempt a comprehensive essay on the teaching of mathematics. Too many educationalists, who have tried to do just this, have fallen into the trap of imagining that there is necessarily a best way of teaching particular aspects of mathematics. They fail to appreciate that what is appropriate in one set of circumstances may not be appropriate in another, that what suits the personality and intellectual make-up of one teacher may not suit those of another, and, most important of all, that what is intelligible to one particular student may not be equally intelligible to another – even though they are studying in the same class. The history of mathematics is, amongst other things, a library of ideas and methods which have worked with varying degrees of success in the past. For this reason alone, it should be studied seriously by every practising and would-be teacher of mathematics, no matter how elementary the level of teaching to be undertaken.

It is unfortunate that there are some mathematicians who take the view that the history of their subject should be reserved for the specialist. To adopt this position is to rob history of the many significant contributions which it can make to one of the educational problems of contemporary society – how to teach mathematics interestingly and effectively and thus improve the basic numeracy of society at large.

The history of mathematics tells us that there is no quick or immediate road to a proper understanding of numbers. The Greeks were well aware of this over 2000 years ago. They were aware also that it is fatal to attempt to teach people to run before they have learned to walk. To build an edifice of mathematics teaching upon flimsy foundations is to invite disaster. The Greeks were aware also of how important it is that all students should have a thorough grounding in the understanding of numbers. Whilst we would hardly accept the Pythagorean philosphical and religious beliefs in their entirety, there is nevertheless an important truth to be found in their dictum 'all is number'. Modern science and technology have a mathematical basis without which they could not have developed. It is now gradually being appreciated that many other aspects of our culture have significant mathematical content. Mathematics is no longer confined to those things which are obviously quantifiable; it can be related to qualitative categories through set theory and mathematical logic. There is virtually no subject taught in schools or higher institutions of learning which can be said to be devoid of mathematical content.

Numbers remain at the heart of mathematics, and the natural numbers are the basis of numbers. The way in which a pupil first comes to grips with natural numbers is therefore in some sense crucial to his or her whole education – how right the Greeks were to insist on a thorough understanding of numbers as a basis for the study of all philosophy! Yet the natural numbers are first encountered at home or, perhaps, in a nursery school. It is those first formative years that matter and to which ideas obtained from a study of the history of mathematics can make one of their significant contributions. Far from being the domain of the specialist mathematician, therefore, the history of the subject should be studied by every teacher. Some parts of it are inappropriate as areas of study for the non-mathematician because they demand a grasp of mathematics which only the specialist is likely to have. The history of number, however, does not come into this category. There is little in this book which should prove inaccessible even to those who make no claim to be mathematicians. This is not a textbook; it is an account of some of the principal elementary ideas associated with numbers, intended as enrichment for the general reader and as a resource book for the parent and the non-specialist teacher, though some aspects should be of interest to the specialist also.

It seems appropriate that the somewhat haphazard remarks which are to be made here about the teaching of mathematics should be put into some sort of meaningful order. They are therefore grouped together under headings which more or less follow the order of the preceding chapters, though there are fewer sections here than there are chapters in the book. Throughout the discussions, however, it must be remembered that there are certain overriding principles of good teaching which are being taken for granted and hence are not being constantly reiterated. These might perhaps be summarized by saying that creative activity and the satisfaction which it brings are just as essential in learning mathematics as in learning any other subject – even the arts! Teaching, divorced from everyday experience, is unlikely to be successful, and teaching which does not exploit the intuition of the pupils, is even more unlikely to be successful. Discovery and a sense of achievement are the secrets of motivation in learning. Mathematics, however elementary, is not a dead collection of facts and methods – it is an exciting life of practical and intellectual adventure and must be presented as such from the moment in which a child first encounters 'one, two, three, . . .'.

Words and Symbols

We have said that there are dangers in attempting to extrapolate back into prehistory using the psychological insights gained from the classroom. The reverse does not, however, necessarily apply. A study of the historical evidence relevant to ancient times can suggest ways in which awareness of numbers may first be encountered. We saw that adjectival forms of number words suggest that the awareness of numbers in the abstract was quite a late development, and we should beware of attempting to impose such an abstract awareness on our children at too early an age. Whilst the step from two parents, two feet, two hands, two pencils, and so on, to two as an abstract concept may now be an obvious and acceptable one, we should remember that for a long period of man's history it was not at all obvious and that it was possible for quite developed mathematical skills to exist without that step being made. To a small child gradually building up his vocabulary, the word 'two' may indeed seem to be of the same kind as 'big', 'small', 'black', 'hard', rather than 'man', 'mother', 'pencil', 'desk'.

A basic difficulty arises when children are first presented with numerals. In many books the symbols 1, 2, 3, . . . are presented as if they were equivalent to letters of the alphabet. There is however a fundamental difference here which must not be ignored. A picture of a cat, together with the letter 'c' or the word 'cat' is not conveying the same sort of information as a picture of two cats together with the numeral '2' or the word 'two'. The letter 'c' stands for a sound – the sound with which the spoken name of the animal pictured begins. The word 'cat' is the written form of that name. The numeral '2', on the other hand, is not the name of the animals pictured, spoken or written, nor is it related directly to that name. It is a property in some way possessed by the two cats by virtue of their being two of them. It does not stand for a sound with which any word begins. It is not the written form of a word in the way that 'cat' is; it is a symbol for a word bearing no obvious connection with the word which it symbolizes. Worse still, the concept which the word symbolizes is an abstract one, and the route to it is the same route that has to be followed in order to understand concepts such as redness through experiencing red pencils, red clothes, red books, red apples, and so on. The route is adjectival though the eventual goal is not. There are thus dangers inherent in those children's books where we find pictures associated with numerals presented in the same way as pictures associated with letters or words, and these dangers are occurring at the first crucial formative stages of a child's encounter with numerals and number words. Indeed, we may well query the introduction of numerals in a child's first encounter with numbers. Numerals do not really become important until it is necessary to carry out written operations on numbers; it may be that we are introducing them too early and that, in the first instance, we ought to concentrate on number-words.

We have seen the various ways in which counting systems have developed from ancient times and we have noted some of the primitive systems found amongst undeveloped peoples even today. The structure of counting systems and the way in which they can be built up from different bases is not without implications for the classroom. It is possible that too much attention is paid to the decimal numeral system too early and that initial approaches to the names of numbers through exploring systems of number-words would enhance the basic understanding of our decimal system and make it easier to introduce later the non-decimal systems which so many authors of modern school books seem to favour. The flexibility which comes from different systems of number-words has much more historical justification than that obtained from artificial changes in the base of the system of numerals. Sooner or later, in this computer age, children should be taught that base 10 is not the only base for written numbers and that, in appropriate circumstances, base 2 has many advantages to offer. Base 60 should also be taught, if only because of its continued relevance to geographical, navigational and astronomical mensuration. However, there is no point in teaching a pure base-60 system – such a system has never existed. Equally, there seems little point in asking children to convert to and perform calculations in unrealistic bases such as 4, or 6, History provides us with systems of both spoken and written numbers whose principles are adequate for all the needs of the classroom; there is no point in inventing fictitious systems simply for the sake of presenting the unfamiliar.

Many of these early systems of numerals were closely associated with units of measurement. This brings us again to the point that the idea of numbers in the abstract is a late idea historically and should probably therefore pedagogically be deferred as long as possible. The concrete should always precede the abstract – abstract concepts are very difficult to assimilate unless there have been plenty of concrete examples with which the pupil has become familiar.

There is no reason why children should not be invited to create their own systems of words and symbols for numbers. This can be an amusing and creative activity and lead to an awareness of the structure of number systems which is otherwise very difficult to teach. However, there is no reason why they should be forced to use pencil and paper. Plasticine and objects which can make indentations of various shapes are adequate – after all, this is effectively how the Babylonians wrote their numbers! Having invented their own systems, pupils might then be invited to test them out on their schoolfellows. It will become apparent that what is meaningful to one is not necessarily meaningful to another.

The communication of numbers is probably the simplest and most basic form of communication possible. This is why it has been suggested that we should attempt to communicate numbers rather than words to other intelligent beings in space – assuming that such beings exist! The history of mathematics tells us just how universal the tallying principle is for communicating numbers; it is when local conventional symbols are used that the ability to communicate easily is lost. Children should be encouraged to invent systems which are easily communicable; they will then come to understand just where and how difficulties in communication arise. They should, however, be encouraged to think other than solely in terms of written symbols. Here again, history suggests that experiments in the classroom with finger-numbers and with knot-numbers may be educationally rewarding.

The point which much of what has been said so far seem to be suggesting is that a great deal of thought needs to be devoted to how we devise a suitable environment and how we develop appropriate teaching methods for the very early encounter with numbers. Much can be done before any question of performing calculations arises. It may be that by rushing to calculation too soon, some teachers are doing irreparable damage by a premature curtailment of that first exploratory period of encounter with number-words and symbols which can be so valuable if properly exploited, and which may well decide children's attitudes towards numbers for the rest of their lives.

Elementary Calculation

We have suggested that it may be the case that children are pressured into performing calculations too quickly. When we study the various methods of carrying out arithmetic calculations that have been popular in the past, the crucial role played by a proper understanding of the number system being used becomes very apparent. Early calculations which children are asked to carry out should therefore be directly associated with the properties of the decimal number system, both spoken and written, and designed to reinforce a child's understanding of it.

The fundamental arithmetic operation is addition. This fact will only be understood if other operations are first introduced in their additive form. Children do not usually have special difficulties with addition, apart from the problem of carries. It may be worth the trouble of following the example set by early Hindu methods of addition where each place is added independently and the carries are added subsequently. This has the advantage of permitting individual children to add in the way which comes most naturally to them, starting either on the right or on the left. Of course, there are advantages to be gained by starting addition with the lowest place, but this does not necessarily imply that every child should be forced to add in this way from the start.

There is no reason why the teaching of addition on the fingers should be treated with disdain. Finger-calculation was virtually a universal practice until the general acceptance of the Hindu–Arabic numerals and the calculation methods associated with them. However, we ought to note that Leonardo of Pisa, one of the principal advocates of calculating with these numerals, wrote:

> It behooves those who would become adept and expert in the art of computation to learn to count on the fingers.

When we come to consider multiplication, we find that he also wrote:

> Multiplication on the fingers must be practised constantly.

Children do not have pen and paper all the time. Calculation on the fingers can be very effective. It can be both interesting and entertaining for younger children. Also, it is immediately obvious to the eye that, in adding (say) three fingers and five fingers, the answer does not depend on the order in which we take the three and the five. It is less immediately apparent with written calculations particularly if complete tables up to $10 + 10$ and 10×10 have been learned by heart. (It may be that the full tables ought never to be pinned up on a classroom wall!) Every teacher in a junior school must have come across the child who can give the correct answer to $8 + 6$ and 6×9 but has difficulty in calculating $6 + 8$ and 9×6. Although we would not dream of using the words 'commutative property' in teaching a six-year old addition or multiplication, it is important that the irrelevance of the order of terms in addition and multiplication should be appreciated from the start, just as that the crucial role of order be understood when it comes to subtraction and division. Here again, we would be following the historical process – history teaches us that it is not necessary to have technical terms available in our vocabulary before being able to appreciate the elementary properties of numbers and operations.

There have probably been more different methods of multiplication than any other of the basic operations on numbers. Although we have looked at a few of these, there are many others which we were not able to discuss. The methods discussed in Chapter 4 do, however, suggest that the way in which multiplication is usually taught in schools today has by no means been the only effective method used. Multiplication can be a significant hurdle for many children, not all of whom ever really manage to surmount it. With long multiplication, there is not only the basic problem of knowledge of tables, there are also the complications of carries, of place, and of the eventual addition. This may well be too much to ask some children to grasp at one and the same time. Here again, history can help us by suggesting methods of setting out calculations in which all these complications do not present themselves at once, or at least not in the same way. Lattice multiplication is a method which immediately suggests itself.

Some children may be able to multiply without difficulty; others may not. It is when difficulties arise that alternative methods ought to be tried with the individual child, not with a view to replacing present-day methods by those of past times, but in order to use those of the past as stepping-stones to those of the present, thus allowing children to face one difficulty at a time.

Some of the ancient methods of calculation are particularly suited to mental arithmetic. They were known and used by people who would not have been able to carry out computations in writing. A number of such examples have been discussed in earlier chapters. Multiplication by a power of two is easily performed mentally by successive doubling – a method fundamental to Egyptian multiplication (and division). Many 'tricks' of mental calculation practised today have been known for centuries. We can immediately think of multiplication by a number close to a power of 10. How many children have been asked laboriously to multiply by 97 instead of multiplying by 100, by 3, and subtracting? 'Trick' methods should not be despised – very often they can introduce principles that otherwise might seem highly abstract and artificial. After all, to apply the principle that

$$(100 - 3)n = 100n - 3n$$

is to make use of the distributive property of multiplication over subtraction – though, of course, we do not need to put this in such technical language in the classroom.

Children ought to be encouraged to check their calculations. Checking by carrying out the inverse operation is the most thorough method. Other checking methods, such as casting out nines or elevens, so popular for many centuries, should not be ignored, however. In addition to reducing the amount of work involved in checking, they raise a number of mathematical principles, perhaps best introduced in just this context where they have an obvious and useful application.

The question of aids to calculation in the classroom is a highly topical one, especially in view of the wide availability of pocket calculators. There is much to be said for introducing these reasonably early on for calculations which are incidental to the particular problems being studied. They are no substitute, however, for a basic ability in and understanding of the elementary operations of arithmetic. Their value in teaching mathematics lies in their ability to take much of the drudgery out of calculation, and thus enable teacher and pupil to concentrate on the problem-solving aspects of the work in hand. They make the introduction of a wide range of practical problems possible, many of which would otherwise have to be excluded because of the length of the calculation involved. Practical problems have a habit of not presenting themselves with convenient figures which divide exactly! Calculators also work to a fixed maximum number of places, and are extremely useful pedagogically for introducing decimal fractions and also the principles and problems arising out of approximation.

Although pocket calculators are now being used more extensively, we must not forget that they were preceded by other aids to calculation such as the slide-rule and the abacus. Indeed, the abacus is still very effectively used in certain parts of the world, especially in the Far East. The typical abacus to be found in toy shops in the West, however, differs considerably from those in use in China and Japan, which, as we have seen, are divided into two regions, a region for fives and a region for units. These Eastern abaci may well be of value in the early stages of mathematical education since they relate directly to counting on the fingers where there is a natural five-unit – the hand. The high skill in abacus calculation developed in the Far East has been made possible in part by the particular form of abacus used. Eastern abaci are not readily available in the West, but they are well within the manufacturing capabilities of many school workshops. Indeed, the manufacture of such abaci for use in the lower forms of a school would be a rewarding exercise for the more senior handicraft classes.

It is not necessary to have ready-made abaci to hand in order to make pedagogical use of the principles of abacus calculation. The forerunner of the modern abacus was the counting board, used extensively in antiquity. Young children can make their own counting boards simply by drawing on paper or card. The counting-board is closely associated with the place-value principle, though it was widely used by peoples who had other and less convenient systems of numerals. Calculation on counting boards using beans, counters or pebbles, or any other suitable form of counters, is a very valuable exercise for young children and assists with the understanding of such things as carries, 'borrowing' in subtraction, and the place-value principle itself, all of which are less easily appreciated from books or the blackboard. Most important of all, counting-board calculation involves a comparatively exciting activity compared with the use of pencil and paper. Above all else in mathematics teaching, it is necessary to keep up an impetus of excitement and discovery. This is particularly true of the primary school where concentration can so easily flag if the task at hand is made to appear dull and uninteresting.

Although the slide rule has largely given way to the pocket calculator for obvious reasons of speed and accuracy, the pedagogical value of making one's own rules of various kinds remains. Napier's rods are easily made by schoolchildren and the historical route by which the modern slide rule evolved can be followed through with advantage. If lattice multiplication has been taught at some stage, the principles of Napier's rods will already be understood. Many different kinds of graduated rule can be experimented with, including those with arithmetic scales (to be used for addition and subtraction) and those with geometric scales (for multiplication and division). It is not necessary to mention the word 'logarithm'; it is sufficient to introduce arithmetic and geometric series and to utilize the rules themselves in order to introduce the principles of logarithms. Here again is an opportunity to do something practical and concrete before introducing theoretical material. Indeed it is not even necessary to introduce series as such. As is so often the case, with a little forethought a great deal of technical jargon can be postponed. More often than we are aware, it is the jargon which is the hurdle that a student cannot overcome rather than the mathematical concepts being introduced. After all, virtually every profession has realized the advantage of inventing sufficient jargon to ensure that it is held in respect by the layman! In teaching, however, our aim should be to pull down barriers, not erect them.

Algebra

One of the great mistakes in the teaching of algebra is to present it as if it were a subject unrelated to arithmetic. It is true that in modern algebra the mathematician is concerned with the manipulation of symbols and with operations which do not necessarily correspond either to numbers or to the operations of arithmetic.

Until the nineteenth century, this approach to algebra was unknown. It is a highly sophisticated approach very largely unsuited to schools, although a number of teaching experiments have tried to introduce abstract manipulation of symbols to pupils at quite an early age. It is doubtful if many educational advances have been achieved by this.

The solution of equations arises from the application of algebraic methods to problems. Just as crucial as the algebra itself is the process by which we proceed from the original statement of the problem to its algebraic formulation. It is most important that this is not overlooked.

Algebra, as taught generally in schools, is not divorced from numbers. The given constants and the unknowns are just numbers and hence behave exactly like numbers. Many schoolchildren, however, arrive at the conclusion that algebra and arithmetic are different subjects. The introduction of letters of the alphabet somehow transports them away from the numbers with which they are at last becoming familiar into new, strange, and therefore forbidding territory. Although it may be a little time-consuming, it is well worth while teaching a little elementary algebra without using symbols other than numerals. After all, the history of mathematics teaches us that even cubic and quartic equations were successfully solved without the benefit of modern notation. The point is, of course, not to discard modern notation for the sake of it – no one should introduce an element of inconvenience without good reason. The use of words instead of symbols can illustrate how close the formulations are to the original statements of the problems.

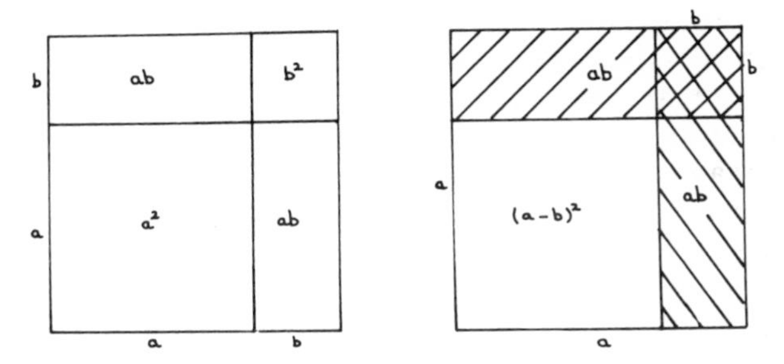

Geometrical representation of expansion of $(a \pm b)^2$

Although the Greeks were extremely subtle in overcoming difficulties associated with the discovery of irrationals, the resulting divorce between numbers and geometry had many unfortunate consequences. The contemporary decline in interest in geometry and its gradual disappearance from school curricula is something which should be deplored by teachers and users of mathematics alike. Geometry is the most visual of the mathematical disciplines. It is not in principle divorced from numbers, and hence neither is it divorced from algebra.

Many a pupil's understanding of algebraic proofs would be considerably reinforced by the introduction of the kind of visual geometrical proofs which were the hallmark of Greek mathematics and to some extent of Arab mathematics also. This is not to suggest that we return to the boredom of the traditional teaching of Euclid as it was presented in schools up to the middle of the present century. It is to suggest, however, that where a geometric proof is clear and immediate, as is the case with the geometric representation of many algebraic identities such as $(a \pm b)^2 = a^2 \pm 2ab + b^2$, the geometry should not be forgotten. The Greeks were some of the greatest teachers of all time, particularly of mathematics. It is not without significance that geometric algebra was in many ways the greatest achievement of Greek mathematics.

Many scholars go right through the period of their mathematical education without being introduced to indeterminate (Diophantine) problems. This should seem somewhat strange to us when we reflect on how many real problems of life require solutions which are meaningless unless they are whole numbers. The absence of Diophantine analysis from school curricula is difficult to justify. The methods required for the solution of such problems are not beyond the capabilities of schoolchildren. It is true that the problems often involve tedious calculations, but this should not be used as an excuse for ignoring the valuable principles of this analysis. Diophantus himself was usually content with finding just one solution to a Diophantine problem. But, once one solution has been obtained, the general solution can readily be obtained from it. It may well be that, because time is limited and has to be allocated on some basis of priorities, Diophantine analysis has received a low priority which ensures its exclusion from the usually highly overcrowded school curriculum. We might perhaps argue, however, that, given the time spent on the three methods of solution of quadratic equations – two of which are seldom used in practice – the priorities in algebra deserve some reconsideration. If it is accepted that too much time and effort is spent on quadratic equations, any room found as a result may be more profitably spent in other ways than on problems requiring natural number solutions. Noting the total omission of such problems does, however, provide motivation for a close look at the real value of some of the algebra to which so much time and effort is often devoted.

Numbers and Life

We have laid considerable stress on the importance of approaching the abstract concept of numbers through the concrete problems of everyday experience. However successful we may be in doing this in our teaching, the question of what numbers are and how they relate to the real world is one which lies at the heart of many of the difficulties experienced in mathematics teaching. We cannot expect to present at pre-university level the more abstract approaches to the definition of the natural numbers which were a feature of the late nineteenth and early twentieth century. Indeed, it is doubtful if such ideas ought to be introduced in the universities at the undergraduate level. We have therefore to accept the situation that the natural numbers are 'given' to us; we do not need to define them. Our pupils will discover ways in which numbers relate to the real world for themselves – all we have to do is to provide the environment in which this can happen easily and effectively.

We should be very careful not to inhibit the obvious ways in which the natural numbers can be extended to the integers and the integers to the rational numbers. At no point should we claim, for example, that three cannot be taken away from two. To say this to a pupil is to erect a barrier to the instinctive exploration of numbers. It is only if we are specifically confining ourselves in an artificial way to working with the positive integers that it becomes true that we cannot take three from two. If we are merely starting from positive numbers, then something exciting happens – we encounter a new kind of number, not positive but negative. Similarly, if we begin with just the integers, it is not long before division ensures our encounter with yet another form of number, the fraction. This is, in fact, the best way to allow fractions to arise. It establishes a kind of pattern of discovery – new things are discovered as a result of operations on familiar things. It is a process which continues into the realms of advanced mathematics, and it is unwise to stifle it at any stage, however humble.

What is a fraction? Again, a study of the history of mathematics suggests to us that there are other ways of explaining fractions than as parts of a whole. Whilst the idea of cutting a birthday cake may prove to be an attractive one in any classroom, it is not necessarily the best way of introducing fractions. What is passed to each person present at a birthday party is one single piece of cake – not a fifteenth of a cake! If we are given a second helping we think of having two pieces of cake; we would surely never think of having two-fifteenths of the whole cake. Thus automatically, the single piece becomes the new unit and the idea of part of a whole is evaded. It was in just this way that most ancient peoples treated fractional parts – they avoided them by resorting to smaller units so that three-quarters was not regarded as part of a single whole but as three of a new and smaller entity, the quarter. Real life tends to be filled with objects rather than parts of objects – the parts become objects in their own rights. The implications for the teacher are obvious.

Today there is a growing awareness of the pedagogical value of games in the teaching of mathematics. Often, however, we read warnings that games should as far as possible be realistic – the writers have forgotten that almost any game is real to a child provided that he knows how to play it. It is true, of course, that no teaching should appear artificially contrived. With simple games, however, the danger of this is at its minimum. Even very young children can learn a great deal from patterns, and especially from patterns of dots which can be joined in various ways by lines. The games which can be built up from the simple idea of dots and lines are far too numerous to mention here. We must not, however, forget that patterns of dots alone can be a productive source of teaching material. After all, figurate numbers provided the Pythagoreans and neo-Pythagoreans with important theorems about the summing of series. Games involving dots and lines can, however, be used to introduce that most difficult of concepts – the difference between the discrete and the continuous. It is well worth while expending a little effort looking at the games played by the undeveloped peoples of the world. Many of these provide great entertainment while at the same time being directly related to the realities of the social and economic environment of those who play them, be they child or adult. They have not been professionally designed by toy manufacturers on the advice of their market researchers or consultant educational psychologists. They have grown naturally out of the need of the people to learn about life and its basic mathematical and logical demands. It is unfortunate that they are often dismissed as intellectually uninteresting by educationalists steeped in the culture of more advanced societies. They may at first sight seem uninteresting to us as games that we might wish to play, yet at the same time they might well be introduced into the classrooms of relatively advanced societies with great profit. Fortunately, there are now a few firms who have realized that simplicity is no bar to enjoyment, and have therefore turned to the manufacture of games which have been directly inspired by those found in the mud homes and dusty street corners of the simpler cultures of Africa and other continents. These simple games are much closer to reality than the sophisticated and expensive forms of entertainment with which western parents are persuaded by garish advertisements to choke the natural initiative of their children. It is far more rewarding for a child to invent a simple game of his own than to be coerced into playing with some electrical gadget of supposed educational value. The introduction of very simple mathematical games into the classroom is much more likely to stimulate initiative and suggest other games which the children can invent for themselves than the introduction of sophisticated educational technology. Today's adults have grown accustomed to living complicated lives and hence are far too prone to introducing complications into the lives of the young both in the home and in the classroom. A few far-seeing adults are now having the good sense to seek out the simpler life for themselves. For children, life is naturally simple. Let us not complicate it for them sooner than is necessary, least of all in our teaching of mathematics, a subject which already has enough complications of its own.

In this final section of the book, we have not intended to propose specific changes in curricula and methods of teaching at any particular level or in any particular place. Fortunately, much of the teaching of mathematics today, especially in the primary school, has managed to break away from the old straitjacket of 'doing sums'. This has been made possible because of the enlightened approach of many teachers and their acceptance that teaching is much more than a matter of conveying information from those who possess it to those who do not. It is still true in some places, however, that the higher the level of teaching the more traditional the method of presentation used. Much of this is due to the pressures of overcrowded curricula and an outmoded examination system – these are both matters long overdue for much more stringent investigation than has been possible so far.

It is fair to claim that it is a student's understanding of mathematics, above all other subjects, which suffers most from unenlightened teaching methods. What we wish to emphasize here is that the troubles may well stem mainly from the first year or two of a child's encounter with numbers, and that it is probable that it is those who supervise this first encounter who are in the best position to affect the general level of numeracy in society. It all begins with numbers – if children come to fear them or to be bored with them, they will eventually join the ranks of the present majority for whom the word 'mathematics' is guaranteed to bring social conversation to an immediate halt. If, on the other hand, numbers are made a genuine source of adventure and exploration from the beginning, there is a good chance that the level of numeracy in society can be raised significantly. There is a real role here for the history of mathematics – and the history of number in particular – for history emphasizes the diversity of approaches and methods which are possible and frees us from the straitjacket of contemporary fashions in mathematics education. It is, at the same time, both interesting and stimulating in its own right.

Bibliographical Note

We hope that this book will stimulate the reader into making further excursions amongst numbers and exploring other areas of mathematics and, especially, the history of mathematics.

We cannot give a complete list of appropriate books available. Even this would not exhaust all the matters aired in this book, since a few of them are not to be found in books elsewhere. It seems a wise choice to limit the suggestions for further reading to some dozen volumes. There are certain dangers in doing this, not least because it may quite wrongly be inferred that the inclusion of a book in the list automatically means that it is the 'best' book of its kind, or that we necessarily agree with every word in it. Every author approaches a subject individually – neither his style of presentation nor what he has to say will be exactly the same as his colleagues'.

All that should be inferred from the following list is that the books quoted contain much valuable material and are therefore certainly well worth reading. Omission of any particular book must not be taken to imply that it is inferior to those which are included.

Boyer, C. B.: *A History of Mathematics*, Wiley.
Cajori, F.: *A History of Mathematical Notations*, Open Court.
Dantzig, T.: *Number, the Language of Science*, Allen and Unwin.
Datta, B. and Singh, A. N.: *History of Hindu Mathematics*, Asia Publishing House.
Dedron, P. and Itard, J. (trans. Field, J. V.): *Mathematics and Mathematicians*, Open University.
Eves, H.: *An Introduction to the History of Mathematics*, Holt, Rinehart and Winston.
Kline, M.: *Mathematics in Western Culture*, Penguin.
Menninger, K. (trans. Broneer, P.): *Number Words and Number Symbols*, M.I.T.
Popp, W. (trans. Bruckheimer, M.): *History of Mathematics: Topics for Schools*, Open University.
Van der Waerden, B. L. (trans. Dresden, A.): *Science Awakening I*, Woltors Noordhoff.
Wilder, R. L.: *Evolution of Mathematical Concepts*, Open University.
Zaslavsky, C.: *Africa Counts*, Prindle, Weber and Schmidt.

In addition to the twelve books listed above, we must draw the readers' attention to the material prepared for the Open University course on *History of Mathematics*, AM289. This material includes a set of five texts under the general title of 'Counting, Numerals and Calculation'. These are:

Van der Waerden, B. L. and Flegg, G.: *Counting I: Primitive and More Developed Counting Systems*, Open University Unit AM289/N1.
Van der Waerden, B. L. and Flegg, G.: *Counting II: Decimal Number Words; Tallies and Knots.* Open University Unit AM289/N2.
Van der Waerden, B. L. and Folkerts, M.: *Written Numbers*, Open University Unit AM289/N3.
Van der Waerden, B. L.: *Written Fractions*, Open University Unit AM289/N4.
Van der Waerden, B. L. and Neuenschwander, E.: *Methods of Calculation*, Open University Unit AM289/N5.

Further texts prepared for this course by G. Flegg and others should also be of interest, though in a number of cases the material is directly interrelated with the books of Boyer, Dedron and Itard, Kline, and Wilder (included in the main list given above). The whole course is, however, due to be replaced by newly written material in 1986.

Index

(a) *Names Quoted*